Ronald L. Holle · Daile Zhang

Flashes of Brilliance

The Science and Wonder of Arizona Lightning

Ronald L. Holle
Holle Meteorology and Photography
Tucson, AZ, USA

Daile Zhang
Earth System Science Interdisciplinary
Center/Cooperative Institute for Satellite
Earth System Studies
University of Maryland
College Park, MD, USA

ISBN 978-3-031-19878-6   ISBN 978-3-031-19879-3   (eBook)
https://doi.org/10.1007/978-3-031-19879-3

This Springer imprint is published by the registered company Springer Nature Switzerland AG
The registered company address is: Gewerbestrasse 11, 6330 Cham, Switzerland

Flashes of Brilliance

Preface

Why a Book About Arizona Lightning?

What's so special about lightning in this state? Unique factors make this an appealing topic. First, the sheer beauty and awe of lightning in the clear Arizona skies with mountains, sunsets, and saguaros are impressive enough. But where, when, and how much lightning happens in Arizona? How dangerous is it? Why is it where it is, and how do we know? This is actually a global story rather than only for Arizona since modern lightning detection systems were invented in Arizona. We will chronicle these stories in this book.

There is a thread that starts with seemingly impractical basic research at the University of Arizona and other locations in Arizona and elsewhere. What developed from this basic research became a resource that is used every day in a myriad of applications and is seen daily on local television news and weather programs, and all types of personal devices around the United States and the world. How did this happen? Who started it, and how did it spread to this extraordinary extent? Despite this amazing advance, the word "lightning" continues to be misspelled frequently (Fig. 1).

Fig. 1 How to misspell and spell the word lightning (© D. Zhang)

How to Spell Lightning?

How Did the Idea Develop for This Book?

The authors were both in Tucson in the late 2010s, and both answering similar questions about lightning from the public, the media, and in University of Arizona classes. At that time, the first author was with Vaisala, and the second author at the Department of Atmospheric Sciences that later became the Department of Hydrology and Atmospheric Sciences at the university. Through a mutual colleague, Dr. Ken Cummins, we started to compare notes. We were explaining the same basic concepts of how lightning occurs in Arizona and elsewhere and refuting the same myths and misconceptions. As a result, we approached Vaisala, and they agreed to publish a short booklet for the public, students, and media (Holle and Zhang 2017). Cirrus Visual made it into an attractive format here in Tucson—see https://www.vaisala.com/en/system/files/documents/Lightning-Booklet.pdf. It has now been translated into Chinese and Spanish. Both versions can be found at https://lightningdev.umd.edu/aert/Safety.html.

However, we had too much material left over from the booklet. It is actually more difficult to write a short explanation of a lightning topic than it is to describe all of the angles and twists that were reduced to a one-page brief summary in the booklet.

At about the same time, the first author was finishing a Springer Press book titled "Reducing Lightning Injuries Worldwide" (https://link.springer.com/book/10.1007/978-3-319-77563-0) with Dr. Mary Ann Cooper that emphasized medical and safety issues related to lightning for a global audience, including our African colleagues (Cooper and Holle 2018). We then approached Springer Press with the idea that another book oriented to the special situation of Arizona could be developed. We combined the interest from Springer from the previous book and the wealth of ideas that came from the booklet and prepared the present book.

This book is intended to satisfy the curiosity of the public about lightning, especially in Arizona. However, even though the book mainly focuses on Arizona lightning, readers from all over the world can learn about this fascinating phenomenon. It also provides a resource for schools at all levels, as a resource available at libraries, as well as for the media. In addition, close to our career paths, the book also includes a review of how the globally important invention of real-time lightning detection has changed an entire scientific discipline. That research thread is winding down at the University of Arizona and diminishing at Vaisala in Tucson. As a result, it is time to describe this continuous line of effort over a long period, since both of us have been closely connected with this endeavor for an extended time. In addition, lightning has been declared an Essential Climate Variable (https://public.wmo.int/en/pro grammes/global-climate-observing-system/essential-climate-variables) by the World Meteorological Organization, an agency of the United Nations (Aich et al., 2018). As a result, the need for better lightning measurements and understanding is growing steadily on a global scale.

While writing this book, we were saddened to hear of Dr. Changming Guo's passing. He was a good friend and colleague to the first author since he visited the NOAA Research Laboratories in Boulder, Colorado. He was probably the first Chinese scientist who came to the University of Arizona and studied lightning physics and lightning detection (Sect. 7.1.2). The second author, who is also from China, is the last lightning student before the atmospheric electricity program at the University of Arizona was closed. Her college teachers were Changming's "grand-students." Through such connections, passion, and dedication, lightning science and detection techniques have been passed through generations and will continue. Although never met, Changming is a name that inspired her to pursuit her dream in the lightning and atmospheric electricity field and will always be remembered.

Enough background…we have so much to talk about! Each chapter of this book has some questions and references that will help the readers explore the topics in more depth. The lightning community is global but there has been a strong presence in Arizona because of lightning's beauty, its frequency, and its mysteries. Off we go!

Tucson, USA   Ronald L. Holle
College Park, USA   Daile Zhang

References

Aich V, Holzworth R, Goodman SJ et al (2018) Lightning: A new essential climate variable. Eos, 99, https://doi.org/10.1029/2018EO104583

Cooper MA, Holle RL (2018) Reducing lightning injuries worldwide. Springer Natural Hazards, New York, 233 pp. (https://link.springer.com/book/10.1007/978-3-319-77563-0)

Holle RL, Zhang D (2017) So you think you know lightning: A collection of electrifying fast facts. Vaisala, Inc., 64 pp. (https://www.vaisala.com/en/system/files/documents/Lightning-Booklet.pdf and https://lightningdev.umd.edu/aert/Safety.html)

Acknowledgments

The authors of this book have benefited from interactions with colleagues at universities, government agencies, and private companies with interests in lightning around the world. The lightning research community is very small, and often interacts with each other globally. They are passionate about the topic as a lifelong career so that lasting professional and personal relationships are formed. We cannot list them all, but several people deserve special mention with connections to Arizona. The list is incomplete and forgive us for missing colleagues who are obvious connections.

We recognize with great respect the core event that made the topics and contents of this book possible. In the middle 1970s, a group of faculty and staff at the University of Arizona built on the basic research up to that time and developed the first real-time lightning detection system. In alphabetical order, they are Dr. E. Philip Krider, who continues as Professor Emeritus in the Department of Hydrology and Atmospheric Sciences at the University of Arizona. Also of importance at that critical moment was Mr. Carl Noggle, who continues to examine various features of lightning and related phenomena in Tucson. Another critical member was Dr. Richard Orville, now at Texas A&M University, who matured the detection technology into the National Lightning Detection Network (NLDN). In many respects, an important individual throughout this entire history is Dr. Martin Uman, now at the University of Florida. Somewhat later but equally important are the continuing contributions of Dr. Ken Cummins, formerly at Vaisala and now at the University of Arizona. To them we owe respect and gratitude for a technology that has changed the way we have looked at lightning ever since. The authors have had the privilege of working with this group in varying situations. The first author has known all of them since the first direction finders were installed in Florida in the late 1970s. The second author had Drs. Cummins and Krider on her Ph.D. committee at the Department of Hydrology and Atmospheric Sciences, so our connections to the foundation of this direction of lightning research run deep.

The thread of connections from this University of Arizona invention of real-time lightning detection led to the establishment of the NLDN Network Operations Center that was in Tucson from the middle 1980s through late 2019. In the early days, there were Michael Maier, Dr. Burt Pifer, Leon Byerly, John Cramer, and others before the network became operational. In particular, there are major participants who made Lightning Location and Protection (LLP), Global Atmospherics Inc. (GAI), and now Vaisala become such successful operations. They also provided firm scientific guidance in the field of lightning detection in addition to the original founders. These include, in alphabetical order, Dr. Ken Cummins, Dr. Martin Murphy (a former Ph.D. student of Dr. Krider), Dr. Amitabh Nag (now at Los Alamos National Laboratory), and Dr. Ryan Said. These scientists are well known globally throughout the lightning community and have provided the foundation for us to develop the contents of this book. We also recognize Mr. William Brooks, who for a long time was a data analyst with Vaisala in Tucson and contributed many of the maps and figures in Chap. 4. We also recognize the faculty and staff at the University of Arizona in the Department of Atmospheric Sciences and the Institute of Atmospheric Physics for generously providing us with references and other help that were needed to complete the review of these developments.

The first author of this book is also grateful to the many other staff members of LLP, GAI, and Vaisala in Tucson. There are too many people to thank personally, but it has been, and continues to be, an amazing journey. Professional and personal relationships started to be formed in the late 1970s with these innovators as lightning detection was developed, as documented in Chaps. 6 and 7. In fact, a sizable group, mainly from LLP and GAI days, meets occasionally for lunch at a favorite Mexican restaurant in Tucson, decades later!

Additional personal connections in recent years within Arizona have been Dr. Randy Cerveny at Arizona State University, and Mr. Ken Drozd and others at the Tucson and Phoenix National Weather Service offices. We also personally benefit in many ways from our close friend Dr. Mary Ann Cooper in River Forest, Illinois, with whom the first author wrote the previous

Springer book on medical and safety issues of lightning that was oriented to some extent to Africa.

We also thank the photographers for working with us on the pictures that are included in this book. Shirley Holle contributed several photos in Chap. 3, and the lightning photographers allowed us to include their special photos. In alphabetical order, they are Lori Grace Bailey, Gary Ladd, Greg McGown, Mike Olbinski, and David Rankin as well as the safety image from Dr. Roni Grad in Chap. 5.2.3. Outside of Arizona, we have had a long-standing connection with Kenneth Langford in Colorado about lightning, lightning photography, and its video analysis. If you have the inclination, visit these photographers' websites, and patronize their businesses; they have spent countless hours collecting these photos!

Unexpectedly, two cartoons were provided to us during the course of writing the book. The monsoon depiction in Fig. 2.5 is by Mr. David Fitzsimmons, an editorial cartoonist and columnist at the *Arizona Daily Star* newspaper in Tucson. The lightning safety drawing for Arizona in Fig. 5.4 was provided by Ms. Wei Xu, a former meteorologist, and a current freelance cartoon artist.

Special mention is due to the members of the National Lightning Safety Council (http://lightningsafetycouncil.org/LSC-Home.html) in addition to us. In alphabetical order, they are Kristin Calhoun, Donna Franklin, John Jensenius, Kimberly Loehr, William Roeder, Katie Flanagan, Chris Schultz, and Chris Vagasky. All of these have been colleagues, some for many years, who bring expertise to lightning safety and education efforts.

And the first author most gratefully thanks his wife, Shirley, and children Eric, Laura, and Paul for their untiring acceptance that preparing a book of this type is an extra duty and arranging quiet time to carry it out! The second author expresses the deepest appreciation to her parents for their endless love, understanding, and support. There are no shortcuts to assembling this material, and family support is always critical!

The starting point for this book was the content of the booklet "So You Think You Know Lightning" (https://www.vaisala.com/en/system/files/documents/Lightning-Booklet.pdf and https://lightningdev.umd.edu/aert/Safety.html) that was funded by Vaisala, championed by Ms. Melanie Scott of Vaisala a few years ago, and made into a highly readable format at Cirrus Visual in Tucson by Ms. Geri Rosen. We had so much material left over that we knew we should write a book! How can the topic of lightning be dull...and we hope we have projected that!

Contents

Abbreviations and Definitions

Lightning

CG flash   Cloud-to-ground flash. A cloud-to-ground lightning flash has
                    one or more return strokes.
CG stroke   Cloud-to-ground stroke. One of the components of a cloud-to-
                    ground flash.
CG return stroke   Cloud-to-ground return stroke. The intense luminosity that
                    propagates upward from earth to cloud base in the last phase
                    of each lightning CG stroke of a cloud-to-ground discharge
                    (https://glossary.ametsoc.org/wiki/Return_stroke).
GLD360   Global Lightning Dataset 360 lightning detection network
                    owned and operated by Vaisala.
IC   In-cloud lightning. About four times as many lightning events
                    occur in clouds without reaching the ground. Also called cloud
                    pulses.
LCC   Long continuing current occurs when current continues to flow
                    between the individual cloud-to-ground strokes of a cloud-to-
                    ground flash.
NLDN   National Lightning Detection Network owned and operated by
                    Vaisala.
Total lightning   Sum of CG and IC data.

Human Impacts

Casualty	The sum of deaths and injuries.
Death; fatality	A person killed by lightning.
Injury	A person injured but not killed by lightning, as used in this book. Note that the medical community often uses the general term injury for both those killed and those who survived.

Organizations

ACLENet	African Centres for Lightning and Electromagnetics Network. A pan-African Network of Centres dedicated to decreasing deaths, injuries, and property damage from lightning across Africa (https://aclenet.org).
LSESSI	Lightning Strike and Electric Shock Survivors International. A non-profit organization in the United States dedicated to survivors, their families, and other interested parties (https://www.lightning-strike.org/).
NLSC	The National Lightning Safety Council is a multi-disciplinary group of United States participants who are dedicated to continuing the long-term downward trend in the number of lightning deaths and injuries in the U.S. and globally (http://lightningsafetycouncil.org/).
SALNet	South Asian Lightning Network. An organization headquartered in Nepal dedicated to improving lightning knowledge and reducing lightning casualties in South Asia (https://salnet.asia/).

1

The Scientific Basics of Lightning

Abstract Lightning is a source of amazement and concern for virtually
everyone around the world. What is it, how does it happen, why does it have
such seemingly strange and random effects, and how can it act so weirdly?
As a result, stories about how lightning forms and what should be done to
avoid it have been the source of speculation and embellishment everywhere
through the ages. There is an established scientific system about how to look
at lightning, its formation, detection, and how to classify it. This chapter
will describe how lightning forms, and the categories of cumulus clouds
that contain lightning are followed through the cumulus growth cycle. The
chapter continues with a detailed description of accepted and commonly-
used lightning terminology, such as negative and positive cloud-to-ground
flashes and strokes, in-cloud lightning, and heat lightning. Since the mixture
of meteorological factors affecting Arizona lightning does not occur in the
same blend anywhere else, a description is made of the factors that result
in lightning during the monsoon season from July through September.
These factors are daytime turbulence from heating, major elevation changes,
mesoscale convective systems, inverted troughs, as well as hurricanes and
tropical storms. Outside of the monsoon season, large-scale traveling meteorological systems can result in lightning in the state. In addition, lightning
occurrence is associated with rainfall in sometimes complex processes.

© Springer Nature Switzerland AG 2023
R. L. Holle and D. Zhang, *Flashes of Brilliance*,
https://doi.org/10.1007/978-3-031-19879-3_1

1.1   Introduction to the Book

The starting point for this book was our booklet "So You Think You Know Lightning: A collection of electrifying fast facts" (Holle and Zhang 2017). We had so much material that we knew we could write a book! How can the topic of lightning be dull...and we hope we have projected that! While the topic of lightning is global, we focus on Arizona lightning because so much unique research and application development has been concentrated in this one state.

The core event that made the topics and contents of this book possible began in the middle 1970s. At that time, a group of faculty and staff at the University of Arizona (UArizona) built on basic research to develop the first real-time lightning detection system. This technology has changed the way we have looked at lightning globally ever since. The first author has been involved with lightning data since the first antennas were installed in Florida in the late 1970s. The second author had two UArizona faculty members who participated in these developments on her Ph.D. committee, so our connections to the foundation of this lightning research run deep.

In this chapter, we will introduce how lightning is perceived everywhere, and how lightning is formed and categorized with some focus on Arizona. Chapter 2 will mainly focus on lightning in Arizona culture. Subsequent chapters will include examples of lightning photography in Arizona (Chap. 3), the occurrence (Chap. 4) and impacts (Chap. 5) of lightning within the state over the last century, and conclude with a description of the invention of real-time lightning technology (Chap. 6) and associated lightning research at UArizona in Chap. 7.

1.2   Lightning Perspectives of the Public

In every society in every age around the world, lightning is perceived as random, unpredictable, and a source of wonder and fear. Since it can seemingly come from nowhere without warning, is unpredictable, and is capable of causing sudden death or damage, it strikes fear such that stories demand to be made up about how or why or when it happened. To societies who are traditionally tied to agriculture and other outdoor activities, the environment around them is closely watched, so this phenomenon is not surprisingly the source of explanations that are handed down from generation to generation. The circumstances of one lightning death, injury, or damage event are prone to be interpreted as repeatable and become part of the supposed reasons for why it happened at that time and place. As a result, a legend is introduced that says a person was killed or injured by lightning because of what he or she was wearing, holding, doing, and so on. News reports often report facts that are unrelated to the lightning—such as a person had just been to a wedding or was on vacation. The implication is that the victim should never have been killed or injured while in a pleasant situation. Few natural phenomena have the ability of people to make associations that are neither relevant nor helpful.

When the first author was a youth in Indiana, he was told that thunder was from Rip van Winkle bowling in the sky. What? How could this be? Rip Van Winkle is a legendary folk tale character developed by U.S. author Washington Irving over 200 years ago. What is he doing bowling in the sky? Where did this come from? It seemed odd at the time, and yet shows how widespread are the mythology, folk tales, and stories associated with lightning and thunder.

Your lightning stories

What stories have you heard from your family, friends, neighbors, and other people? How do you know if they are true? Why are some of them completely false but still believed? How have they affected your view and reaction to lightning? How can the false stories be dangerous? How can they be refuted or dismissed?

We wrote the booklet "So You Think You Know Lightning" (Holle and Zhang 2017) mentioned above that is available at https://www.vaisala.com/en/system/files/documents/Lightning-Booklet.pdf and https://lightning dev.umd.edu/aert/Safety.html includes additional Spanish and Chinese translation versions. There, we covered some of the more common stories that we have often been asked by the public, media, and university students.

The following are TRUE statements:

- Metal does not attract lightning.
- Soil composition has nothing to do with where lightning happens.
- Lightning can strike the same place twice.
- Lightning does not always strike the tallest object.
- Florida has the most cloud-to-ground flashes per area in the U.S.
- Cell phones do not attract lightning.
- A person can survive being struck by lightning; proper medical care can alleviate some symptoms.
- Rubber tires do not protect cars from lightning.
- Tents and small shelters are not safe from lightning.

- Direct strike is the least likely way to be injured by lightning.
- Do not go under a tree in a thunderstorm.
- Rubber-soled shoes do not save you from lightning.
- Passenger planes are safe from lightning.
- The exact place of where the next lightning is going to strike cannot be predicted by current technology.

Surprised?

Many of these true statements are opposite to widely held knowledge in the U.S. and around the world. And since they can lead to wrong conclusions about how to be safe from lightning, they are not idle topics of curiosity but need to be understood in view of recent knowledge and updates to lightning safety advice.

The following are FALSE statements:

- Florida is the lightning capital of the world.
- A lightning casualty should not be touched.
- I can run away from lightning in a thunderstorm.
- Lightning only strikes the highest peak on a mountain.
- Certain neighborhoods are more prone to lightning than others.
- Some people are more prone to lightning than others.
- Crouching low will prevent someone outside from being hit by lightning.
- Lightning travels steadily in a straight line in a thunderstorm.
- Standing under a power line makes you safe from lightning.

Still surprised?

All of these false statements are heard in the U.S. and many other countries. Don't worry if you thought they were true, but we hope to provide information in this book to correct these common impressions.

For example, one of the vast numbers of lightning safety myths is that in southeast Africa, it is often accepted that wearing red attracts lightning. A formidable global overview of lightning and thunder in mythology and storytelling around the world is provided by Elsom (2015). This is the most comprehensive known summary of how pervasive lightning has been considered globally through the centuries. Included are historic, religious, witchcraft, literature, and art references that make this a fascinating and wide-ranging summary. One can only admire the dedication in developing this multi-disciplinary summary resulting from the passion on the part of the author, now deceased.

1.3   Lightning Formation in Cumulus Clouds

In this section, we provide an overview of cumulus cloud life cycles that involve lightning formation. The emphasis is on how these events occur within Arizona, since this is the focus of the book. Nevertheless, the basic principles are applicable everywhere.

Lightning always begins in clouds whose tops are colder than freezing—in Arizona too! Upward motion, called updrafts, starts in the lower portions of a thunderstorm and then reach up to colder altitudes. The freezing level during the summer monsoon months when most lightning occurs in Arizona is often between 4.2 and 5.2 km (14,000 and 17,000 ft). Since lightning initiates at temperatures between −5 and −15 °C (23 and 5 F), this layer where lightning forms is a kilometer (3,000 ft) higher, or more, at altitudes colder than the freezing level. Occasionally in winter, Arizona lightning forms and starts at lower levels.

Updrafts that result in lightning occur within cumulus clouds. The updraft can be five meters per second (11 miles per hour) and can reach 25 meters per second or more (56 miles per hour) in exceptional cases in Arizona. The upward motion is concentrated in mostly vertical columns of air whose cores may be a kilometer or two across (around a mile). These towers are sometimes only a small portion of the entire cloud, but they are necessary for cloud-to-ground (CG) lightning production.

Let's go through the sequence of cumulus cloud formation and growth until it produces lightning. These are the major stages:

- **Humilis:** Shreds of cumulus are the first to appear, as shown in Fig. 1.1. This mainly flat cloud that is wider than tall is called cumulus humilis.
- **Congestus:** As the updrafts grow stronger, the cloud becomes taller than it is wide. When vertical towers begin to form as shown in Fig. 1.2, these are called congestus and often visible in the afternoon from a long distance away in Arizona as they often form over the mountains. A small amount of lightning may begin at this stage.

Fig. 1.1 Cumulus humilis clouds over Tumacácori National Historical Park in southern Arizona (© R. Holle)

Fig. 1.2 Small cumulus congestus clouds over the Santa Catalina Mountains as seen from Oro Valley, a town located north of Tucson, Arizona (© R. Holle)

- **Cumulonimbus**: As the updraft grows and other cumulus towers combine, a cumulonimbus cloud is formed as shown in Fig. 1.3. An important feature is the upper part of the cloud that is flowing outward in the shape of an anvil. At least some lightning occurs within nearly all cumulonimbus clouds in Arizona.

Fig. 1.3 A single cumulonimbus cloud over the Santa Catalina Mountains as seen from the Arizona campus in Tucson (© D. Zhang)

- **Anvil:** The presence of an anvil indicates that cirrus clouds have formed; cirrus is an ice cloud that can be told by its wispy appearance. When an anvil is seen in the distance, often over the mountains, it is a clear indication that lightning is occurring or about to happen. The direction of movement of the anvil cirrus is determined by the direction and speed of the winds at its altitude and is usually not the same as the wind direction and speed on the ground.
- **Cumulonimbus complex:** As more and more updrafts form, they can develop into a complex as shown in Figs. 1.4 and 1.5. Notice the large number of vertically oriented cumulus towers at small, medium, and large scales. Often, parts of the anvils blowing off the tops of cumulus towers are visible.

The sequence in Fig. 1.6 shows these phases during a 43-min period. Panel (a) shows two tall congestus towers that are not yet raining. Panel (b) shows the consolidation of the towers into a larger mass, then the anvil is apparent in (c) and (d). Figure 1.7 shows a complex of active updraft towers that indicate a significant thunderstorm with lightning is underway and likely to grow

Fig. 1.4 A small cumulonimbus complex over the Santa Catalina Mountains as seen from Oro Valley, Arizona (© R. Holle)

larger and stronger in the coming minutes. This photo was taken in Show Low at an elevation of 1,920 m (6,300 ft) on the Mogollon Rim in east-central Arizona where cloud bases are closer to the ground compared with the lower elevation metropolitan areas of Phoenix and Tucson.

Question

Now that you have seen the main stages of cloud formation leading to one that has lightning, can you tell simply by one look at a cloud whether it is a lightning danger or not? That is not always true, but you should be able to sort out the mediocre flat humilis from the growing cumulonimbus by now!

Timing

What stage from one panel to another in Fig. 1.6 is likely to happen most quickly? Which may be the slowest?

But the sky often does not appear as simple as shown in these examples. Due to the typically dry air at lower levels of the atmosphere in Arizona,

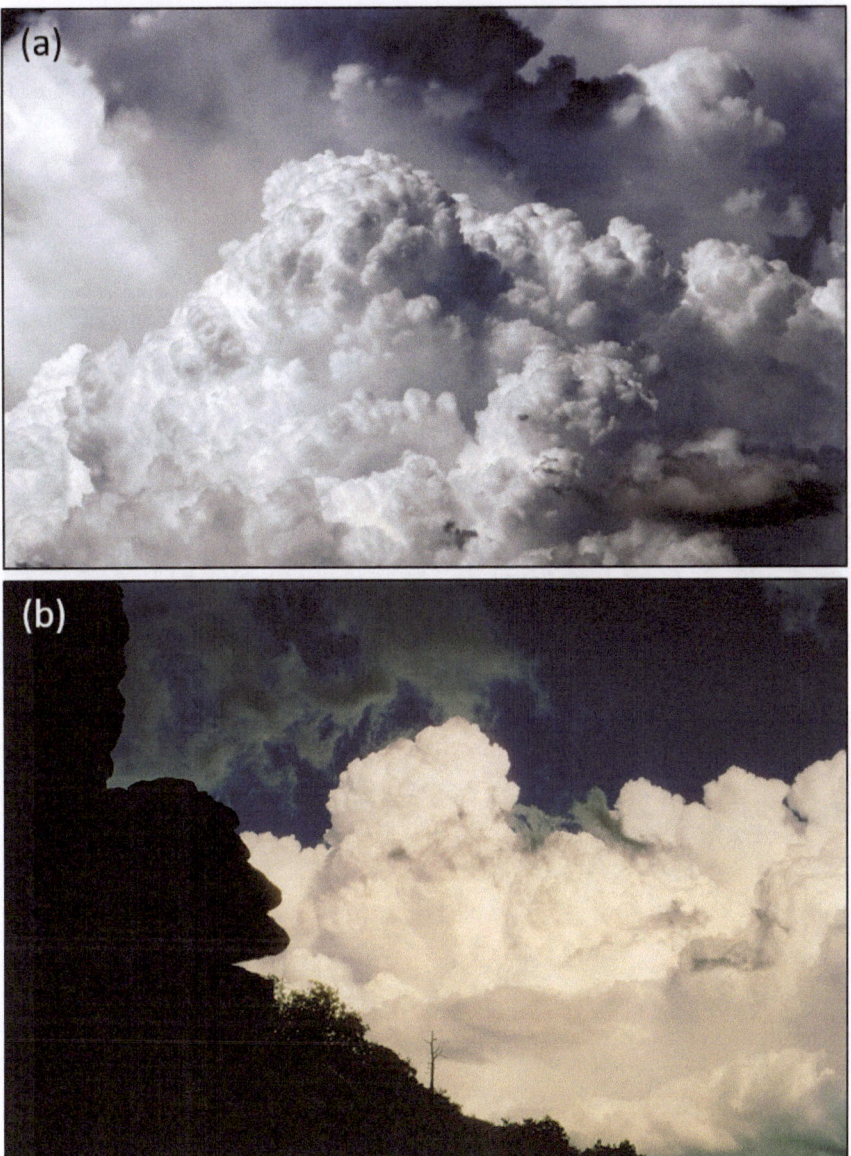

Fig. 1.5 Cumulus towers over Oro Valley (**a**) and the Santa Catalina Mountains (**b**) of southern Arizona (© R. Holle)

rain may not be falling from a well-defined cumulonimbus cloud with lightning (Fig. 1.8). This photo from Klagetoh on the Navajo reservation of northeastern Arizona was taken at an elevation of 1,958 m (6,424 ft). Lightning had occurred in the immediate vicinity with loud thunder 30 s earlier.

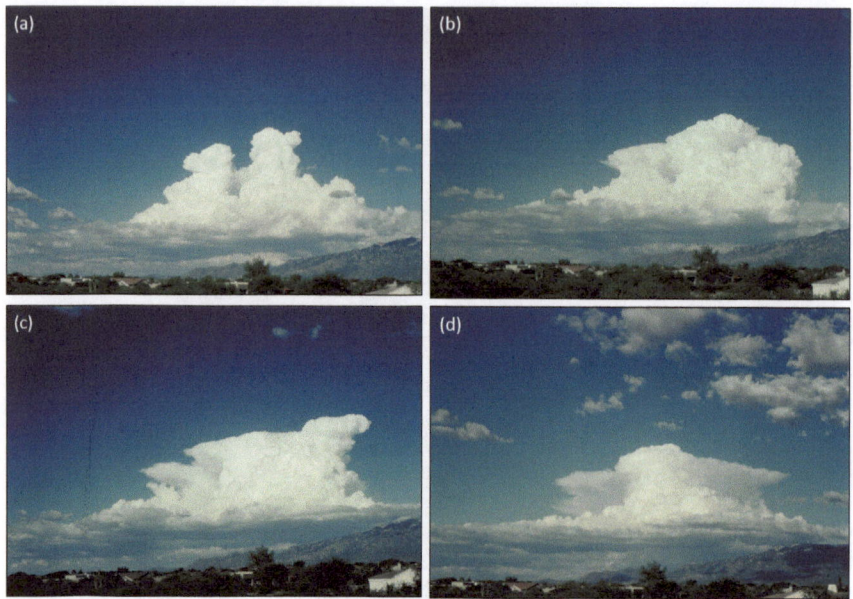

Fig. 1.6 Growth sequence over a period of 43 min of a cumulonimbus near the Santa Catalina Mountains as seen from Oro Valley, Arizona (© R. Holle)

Fig. 1.7 A cumulonimbus complex with numerous active updraft towers near Show Low in east-central Arizona (© R. Holle)

Fig. 1.8 A cumulonimbus overhead that produced lightning 30 s earlier at Klagetoh in northeast Arizona (© R. Holle)

The difficulty of identifying lightning-producing clouds in the arid state of Arizona will be considered in more detail in Sect. 5.2 on lightning safety.

All clouds described here are cumulus whose main features are rounded tops on tall clouds with updrafts. Cumulus clouds are shaped with upper boundaries that appear in puffs, mounds, or towers that have a vertical or sometimes slanted appearance. The other broad type of cloud is stratiform that is generally horizontal. Widely available popular weather books describing cloud types with a plethora of cloud photos are Kahl (1998), Ludlum (2001), and Ludlum et al. (1991, 1995). The World Meteorological Organization (WMO) has an extensive description of all types of clouds and their variations in the International Cloud Atlas at https://cloudatlas.wmo. int/en/home.html.

Cloud Atlas

The WMO of the United Nations recently published a thorough summary and classification of all cloud types. This cloud atlas includes photos of clouds, their descriptions, the associated upper air soundings, and surface weather conditions for various types of clouds that are not seen often or at all in Arizona!

So exactly how does lightning start in the updraft? The updraft starts at altitudes where the air is warmer than freezing, then the upward motion connects the warmer lower levels with a higher layer of air that is colder than freezing. The necessary ingredients for lightning production are known to be an updraft and a mixture of small hail (also called graupel), super-cooled water droplets, and ice particles at temperatures colder than freezing (Saunders 1993; Stolzenburg et al. 1998; Rakov 2016). Graupel is a small irregularly shaped ice particle. Larger ice particles leave the updraft when they become too heavy to be lifted any longer or are thrown out of the cloud to the side. At the same time, the lighter weight particles stay aloft. The difference in size of ice and water particles, as well as wind velocity variations within the cloud, results in collision of particles that carry different charges, which results in the separation of charge, and initiate lightning. The details of how lightning starts with these ingredients within a cumulus life cycle can vary from one time and place to another. The necessity of having ice particles for nearly all lightning production underscores the importance of paying attention to the wispy, diffuse appearance at the top of cumulonimbus clouds in Figs. 1.3, 1.4, 1.5 and 1.6.

Summer thunderstorms result from a large temperature difference in the vertical, where the warmest air is at the lower levels and colder air is aloft. In the situation of Arizona, the ground can become very hot while the air aloft remains relatively cool. Air warmer than the surrounding atmosphere begins to rise, and instability develops when the temperature inside the buoyant bubble is warmer than the air around it. The buoyant bubble may start to accelerate as it rises if it becomes still warmer than the surrounding cooler air. That difference increases with height since the warm-air parcel is much warmer than the surroundings and it keeps moving upward. Charge separation starts between the various sizes and types of frozen and liquid particles when the updrafts carry air from lower altitudes to a layer of -5 to -15 °C (23–5 F). When the temperature inside the parcel equals that of the surrounding atmosphere, the warm air stops rising.

What are some unique cloud features in Arizona?

In Arizona, the air is often so dry and clear that you can see many of these cloud features. How far away can you see them from Tucson and Phoenix in July and August? Or is there dust, smoke, or pollution blocking your view part of the time?

1.4   Lightning Terminology

1.4.1   Introduction

Lightning is usually divided into two categories, cloud-to-ground (CG) and in-cloud (IC). Most thunderstorms have a mixture of both, and there are several times as many IC occurrences as CG flashes. Figure 1.9 shows a typical negative CG flash. Note the downward-pointing unsuccessful branches emanating from various parts of the main channel to ground. Figure 1.10 shows a negative CG flash and an IC event at the same time.

> **Definition to remember**
>
> For cloud-to-ground lightning, the most important phrase is the following: **A cloud-to-ground flash has one or more return strokes.**

Let's explore these definitions of lightning in more detail, since they are very often misclassified and misused:

Fig. 1.9 A negative CG flash over Tucson, Arizona as seen at an estimated distance of 25 miles (40 km) from Oro Valley, Arizona (© R. Holle)

Fig. 1.10 CG and IC lightning over the Catalina Mountains as seen from Oro Valley, Arizona (© R. Holle)

- **Cloud-to-ground (CG) lightning**: Lightning that comes from the sky and contacts the earth's surface or an object on the earth's surface.
- **In-cloud (IC) lightning**: Lightning that stays in the sky. Sometimes called an in-cloud pulse. They often are near or attached to CGs aloft.
- **Negative CG flashes**: Usually have multiple strokes. The strokes within a negative CG flash are typically less than a tenth of a second apart in time. They lower negative charge to ground.
- **Positive CG flashes**: Almost always have only one stroke. They lower positive charge to ground.
- **Cloud-to-ground flash**: Includes all of the return strokes in a flash. Remember that "**A cloud-to-ground flash has one or more return strokes.**" A CG flash has a median of about four return strokes (Rakov 2016).
- **Strike**: Imprecise and will be used only for specific situations related to damage. Avoid "strike" as much as possible by not using it to refer to any kind of lightning except a CG that actually contacts the earth's surface or an object on the earth's surface.
- **Bolt**: Undefined and will never be used in this book.

1.4.2 Negative Cloud-to-Ground Flash

Most CG flashes lower negative charge to ground; the ratio of negative flashes based on data from ground-based detection systems is over 90%. The sequence of formation for a negative CG lightning flash is shown in Figs. 1.11 and 1.12. These photos were taken by moving the camera back and forth during the duration of the flash. The sequence is as follows:

- The first stroke in this negative CG flash begins within the cloud aloft.
- The current flows downward toward ground in a faintly visible channel in steps of 30–50 m in length.
- The first stroke often branches near the ground, as shown on the right in Fig. 1.11 but not Fig. 1.12.
- Since branching often occurs only on the first stroke in a negative CG flash, we can tell that the camera was moved from right to left in Fig. 1.11.
- The last 30- to 50-m step that contacts the ground is vertical; the major importance of this vertical channel for lightning detection technology is described in Chap. 6.
- Once the first stroke reaches the ground, a very bright visible light moves upward at about one-third the speed of light.

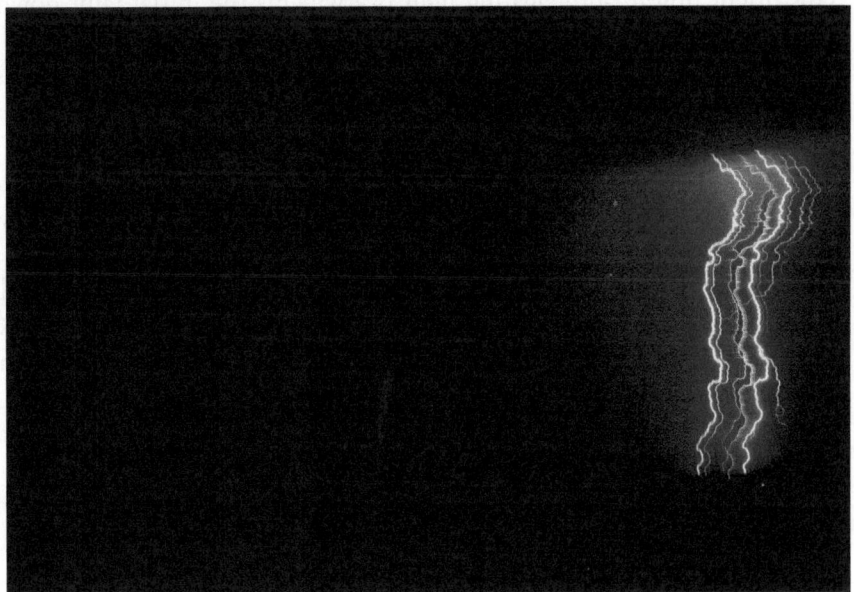

Fig. 1.11 A five-stroke negative CG flash over the Santa Catalina Mountains shown by moving the camera from right to left as seen from Oro Valley, Arizona (© R. Holle)

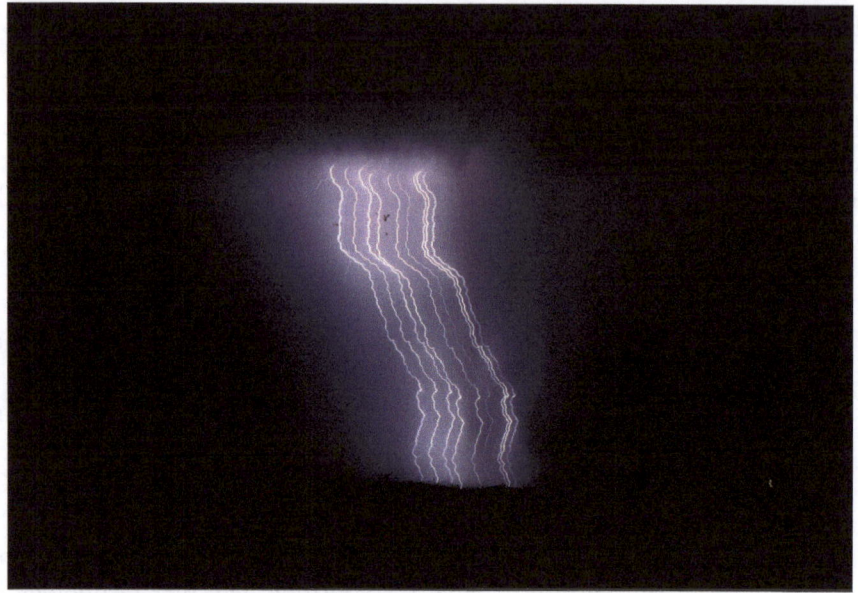

Fig. 1.12 A nine-stroke negative CG flash over the Santa Catalina Mountains shown by moving the camera from right to left as seen from Oro Valley, Arizona (© R. Holle)

- This bright upward-propagating light overwhelms the initial faint downward channel in a still photograph; this is what human eyes see.
- Negative charge is lowered to ground during this process, and this first stroke is typically the strongest in a multiple-stroke CG flash. Sometimes, one of the subsequent strokes will find a new channel to ground.
- If lightning needs to remove (drain) more charge from the cloud, a second return stroke travels down the same channel and lowers more negative charge.
- As a result, the subsequent channels in Figs. 1.11 and 1.12 are exactly parallel and occur within meters of the same location as the first stroke; they only appear separate because of the motion of the camera during the flash—they are traveling in the same channel.
- The light from the second stroke then travels back up the same channel as in the first stroke.
- If charge continues to need to be drained from the cloud, a third, fourth, and sometimes additional strokes will lower charge down the same channel of this negative CG flash while the light goes upward.
- The average number of return strokes in a negative CG flash is around four (Rakov 2016). Unpublished Vaisala maps using the National Lightning Detection Network (NLDN) data show a typical ratio of three to four CG

strokes per negative CG flash for all of Arizona except from the Phoenix area westward into the lower deserts where the ratio is often less than two CG strokes per negative CG flash. A report of a 26-stroke flash is in Uman (2011).

- The photos in Figs. 1.11 and 1.12 show what is sometimes called ribbon lightning. It is normal lightning with multiple return strokes. Since the individual strokes cannot be seen as separate channels in real time, ribbon lightning only appears in photographs. In his popular book, *All about Lightning*, Uman (1986) includes several lightning photos taken in Arizona that are similar to those in Figs. 1.9, 1.10, 1.11 and 1.12.

How wide is lightning?

Arizona was home to one of the earlier studies of the size of the lightning channel (Evans and Walker 1963; Uman 1964). Measurements were made where lightning struck two television towers on the Santa Catalina Mountains just north of Tucson in 1963. Size was found from damage made to 11 fiberglass screens placed on top of the lightning rods. The values were between 2 and 3.5 cm, in other words around an inch, which continues to be the generally accepted value.

1.4.3   Positive Cloud-to-Ground Flash

The formation for a positive CG lightning flash is similar to that of a negative CG flash except for the following differences:

- Positive CG flashes usually have only one return stroke.
- That stroke has no branching.
- Recoil leaders sometimes appear in high-speed videos (Sect. 3.3.3).
- However, since some negative CG flashes also may have only one stroke and no branching, one cannot be certain of the polarity of a flash only by its visual appearance if there is only one stroke visible.

1.4.4   Continuing Current

Sometimes a CG flash will not completely halt its contact with the ground between strokes in a negative CG flash. This phenomenon may also occur after the only stroke in a positive CG flash. When that occurs, continuing current (CC) is said to be applied to objects on the surface of the earth, and that is very likely to be especially dangerous in starting forest fires and causing such damage as outages to power grids due to extreme heating (Sect. 5.3). In such flashes, the current does not disappear entirely in the inter-stroke interval but stays in contact with the ground. Sometimes, imagery (not included here) will include a moderately bright light between the strokes that may be indicated by continuing cloud-top optical energy detected by the satellite sensors, such as the Lightning Imaging Sensor (LIS; Bitzer 2016) and the Geostationary Lightning Mapper (GLM, Fairman and Bitzer 2022).

The percentage of flashes with continuing current varies with polarity. For positive CG flashes, it is about 50% and for negative CG flashes it is about 10% (Rakov 2016). However, since there are about 20 times as many negative CG flashes as positive CG flashes, there are actually more negative CG flashes with continuing current than positive CG flashes. Some portions of the forestry community have overstated the situation by saying that all positive CGs have continuing current, and negatives don't, so continuing current in negative CG flashes is not a concern. That is not correct! Much more in-depth research is underway that should help clarify the situation for users of lightning detection network and satellite data who are interested in which CG flashes have continuing current.

1.4.5   In-Cloud Lightning

There are several times as many IC flashes as CGs. They do not come to ground and do not directly affect people and objects. However, since ICs rarely occur in isolation for very long, their presence is nevertheless a very important indicator that CG flashes could occur in the vicinity. For that reason, ICs are always included in warning systems for CG flashes in order to have a more complete representation of the CG lightning threat (Sect. 5.2).

Most of the time in Arizona, ICs are small as shown in examples from southern Arizona in Fig. 1.13. Since they are within the cloud, much of their path is hidden by water and ice particles, such that only short segments of the entire channels are often visible from the ground. It is likely that some of the IC channels visible in Fig. 1.13 are connected to CG flashes that are outside the view of the camera or are obscured by intervening clouds.

ICs can occur in almost any configuration and very large and long-lasting occurrences outside Arizona are the subject of new research. These have been termed megaflashes (Lyons et al. 2020; Peterson 2021; Peterson and Stano 2021; see more in Sect. 7.3.2). In a study using ground-based lightning detection network data by Lang et al. (2017), these megaflashes were found to last

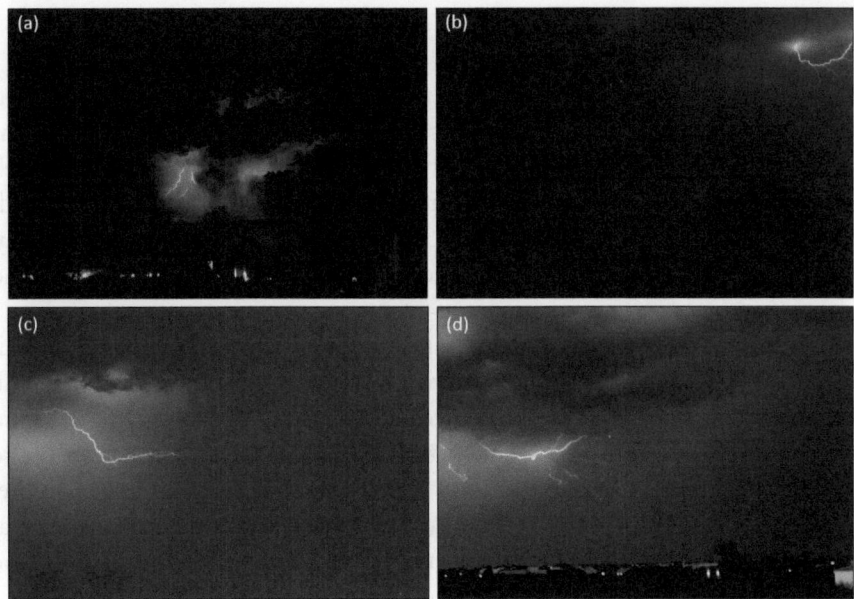

Fig. 1.13 In-cloud lightning as seen from Oro Valley, Arizona (© R. Holle)

in a cloud for as long as 7.74 s and extend as far as 321 km. This study described a flash that was at least 500 km long in Oklahoma and Kansas that was detected by the NLDN as being associated with 17 positive CGs, 23 negative CGs, and 37 ICs that were detected during this long path. In a new study using satellite data and complementary ground-based measurements, Peterson et al. (2022) found still longer flashes extending without any interruption for a remarkable 768 km in length and lasting 17.01 s in time! These extreme events occur in large Mesoscale Convective Systems (MCSs, see more in Sect. 1.5.3) on the plains of the U.S. and elsewhere.

Polarity is assigned to ICs by the following convention for data from the NLDN, as provided by Dr. Ryan Said of Vaisala: "We define the polarity based on the polarity of the received radio impulse, and this corresponds to the direction of charge movement in the cloud. The polarity refers to the direction of charge movement, so that the presence of mostly positive cloud pulses in the NLDN suggests a prevalence of 'canonical' tri-polar charge structure, with current pulses connecting the upper positive with the main negative charge layer."

1.4.6 Heat Lightning

Heat lightning is normal lightning that occurs too far away to hear its thunder. In Fig. 1.14, the event is around 64 km (40 miles) away from the camera. This view is also on page 16 of Holle and Zhang (2017). Light travels at 299,792 km per second (186,282 miles a second); in other words, incredibly fast. In contrast, sound travels 343 meters per second. Said another way, it takes 4.69 s to travel one mile; the velocity varies slightly with pressure and temperature.

As a result, the light reaches you instantaneously. A consequence of this fact is that you have no time whatsoever to react to the light from a flash in terms of safety (Sect. 5.2). Sound takes longer to reach you, and the sound of thunder does not travel too far in the atmosphere. It's usually possible to hear thunder up to several miles away and it's sometimes audible 5–10 miles away in a less populated area. Thunder is usually very hard to hear, only as a low rumble, at 10 miles unless you are in a very quiet location with no distractions such as wind.

The conclusion is that it is easy to see lightning a long way away but not to hear the thunder under most situations at more than 10 miles

Fig. 1.14 Photograph of distant CG lightning to the south as seen from Oro Valley, Arizona. This view is sometimes called heat lightning since the light can be seen but no thunder is heard due to its distance of around 64 km (40 miles) away from the camera (© R. Holle)

away. These differences between light and sound are part of the basis for the 30–30 rule of lightning safety in Sect. 5.2. Another essential outcome of these considerations is "When Thunder Roars, Go Indoors" since any time that you can hear thunder, lightning is close enough to present danger. This phrase is widely used as a first step in lightning safety by the National Lightning Safety Council (http://lightningsafetycouncil.org/LSC-Home.html). Although the 30–30 rule is widely used, extremely large and long flashes occur as mentioned above (in Sect. 1.4.5), so we may need to have a stricter rule for areas with large MCSs.

Question

Which of the ICs in the four panels of Fig. 1.13 seems more likely to be connected to a CG flash?

Heat lightning appears to be more of midwestern and eastern U.S. term. Lightning tens of miles (km) away normally cannot be seen at all during the day anywhere. But at night, lightning far away can be seen if there is minimal light interference. In the midwest and eastern states, storms at night often occur during the summer when the day has been hot, and organized thunderstorms that continue into the evening and nighttime are generally associated with heat, that is, days that were hot. Such nighttime storms may be in long lines on the horizon where only the upper parts of the cumulonimbus clouds with frequent ICs are visible.

In Arizona, the air is often so clear that it is not unusual to see storms a long distance away at night, especially during the summer monsoon season. For example, during the day the tops of thunderstorms over the White Mountains of east-central Arizona can be seen on the horizon up to around 120 km (75 miles) or more away from Phoenix looking east or from Tucson looking northeast. Similarly, the same clear-sky conditions can provide clear views of the tops of thunderstorms looking southward from Tucson to storms over the Sierra Madre Occidental mountains of northern Mexico. While the lightning channels themselves are not visible, the distant cloud tops can light up far away at night.

While we're on the topic of distant cloud tops, there is the related topic of red sprites, blue jets, elves, and pixies. These are electrical phenomena seen over the tops of large thunderstorm complexes but only at a distance that will be described in the next section.

Calculation

How far away is a flash when you hear its thunder 30 s after the light is visible? Does that seem too far away or too close?

1.4.7 Transient Luminous Events, Sprites, Jets, and Ball Lightning

During the last several decades, new middle atmospheric lightning-related phenomena called Transient Luminous Events, sprites, and jets have been identified that are also described in Sect. 7.3.4 for their relevance to Arizona (Lyons 2006; Lyons et al. 2009). They occur over large thunderstorm complexes and have been documented in many areas of the world. They had been there all along, but not observed separately because they are (1) very faint, (2) short-lived, and (3) rare. The underlying cause of these events is usually a large thunderstorm complex, but they can only be seen well at night at a distance, and they mainly occur over the tops of large thunderstorms. The American Meteorological Society (AMS) has precise technical definitions of these terms at https://glossary.ametsoc.org/wiki/Welcome.

Ball lightning? There is still no agreement among the lightning community whether ball lightning exists or not. It's likely a real occurrence, but photographic documentation is elusive (Lyons 2022). A huge range of observations and theories makes it difficult to speculate on what exactly is occurring as well as whether they came from the same phenomenon. But a luminous sphere, sometimes of the size of a basketball, has been reported for centuries (Gasper and Tanner 2022). The AMS website mentioned above also has an entry for ball lightning. The one common feature is that they occur in the vicinity of thunderstorms. The second author of this book has been inspecting meteorological records from China over the last millennium to find reports that could be labeled as ball lightning. It's possible that several types of electrical phenomena occur in the vicinity of normal lightning in thunderstorms that are grouped under the term ball lightning!

1.5   Factors Leading to Lightning in Arizona

The updraft core within a cumulus is essentially the same regardless of how it is formed. The main types of meteorological conditions resulting in Arizona lightning are due to the following factors.

1.5.1 Turbulence Due to Daytime Heating

Over strongly heated land, some upward motions become strong enough during the day to result in updrafts that can reach high enough altitudes for lightning production. Over the open desert or flat high-elevation areas, the locations of such thunderstorms are mostly random. Such storms are likely to have only a few updraft towers, be less than 10 km (6 miles) across, last only an hour, and make lightning for tens of minutes. Such storms may only move a few kilometers (miles) before falling apart.

1.5.2 Major Elevation Changes

There are huge variations in elevation across many parts of Arizona and they strongly affect where updrafts are formed. These vertical motions are forced by air that must change altitude as they encounter mountain ranges, canyons, and locations such as the Mogollon Rim and White Mountains. Often, the most frequent lightning tends to occur on isolated features on the slopes of steep terrain, and not always over the highest peaks (Cummins 2012). The result is that storms forming over the highest peaks may produce only a few CG flashes that are, unfortunately, not perceived at the time to be a major threat to people (Hodanish et al. 2004, 2015). These storms may not move far from the elevation change on the ridge or escarpment that caused them, then the anvil tops often blow off while the updraft dies in the storm below. Note that small hills that are tens of meters high are not adequate to force an updraft strong enough to reach the sub-freezing layer where lightning is formed.

1.5.3 Mesoscale Convective Systems (MCS)

These are prolific lightning producers (Steiger et al. 2007; Dotzek et al. 2005) that cover very large areas—as much as 100,000 square kilometers. They are strongest at night, last for up to 18 h, and have been measured to produce tens of thousands of CG flashes in a single night (Laing and Fritsch 1997).

An MCS that lingered over the Phoenix area for several hours is described by McCollum et al. (1995). While this MCS only had a few thousand CG flashes, it caused major flooding and other damaging winds. MCSs often occur over the higher elevations of northwestern Mexico and sometimes extend into southern Arizona during July through September evenings as in the examples shown in Figs. 1.15 and 1.16. The updrafts in MCSs are

not as purely vertical as during afternoon storms due to heating and elevation changes. Instead, MCSs can develop a large-scale sloping layer that can
persist for hours (Ely et al. 2008). The example in Fig. 1.15 shows that the
largest and most intense convective system in all North America at this time
is over the northern mountains of Sonora and Chihuahua states of Mexico
and extends into southern Arizona. Another MCS is apparent further south
in Mexico. The example in Fig. 1.16 shows an MCS over southern Arizona
extending somewhat into Sonora in the evening, and as is often the case,
not as symmetrical as shown in Fig. 1.15. MCSs usually have a preferred
side where growth is most active, often on the side with the best access to
low-level moisture flowing into it, but each MCS has its own pattern. Such
a growth region is shown by the hard edges in Fig. 1.15 on the west and
southwest sides. Similar MCSs are common in this area during many evening
and nighttime hours during the southwest monsoon months starting in June
in Mexico, and in southern Arizona in July through September (Holle and
Murphy 2015).

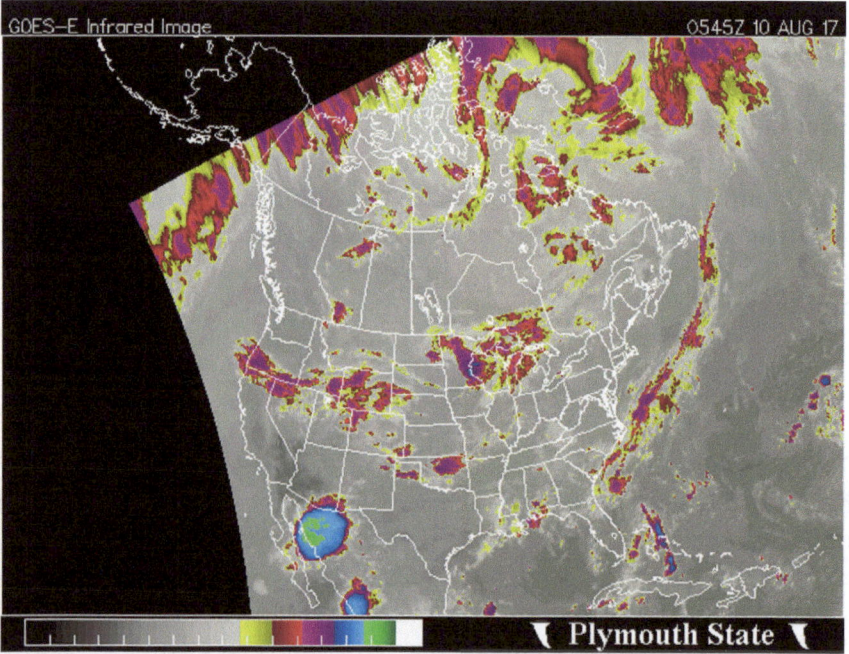

Fig. 1.15 GOES-East satellite view in the infrared of cloud cover over North America
at 10:45 pm Mountain Standard Time on August 09, 2017 [0545 UTC, 10 August]
(*courtesy* Plymouth State University website)

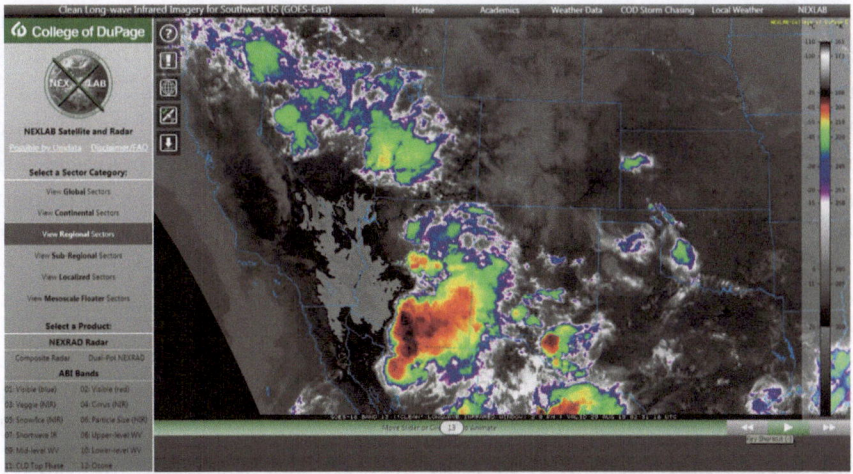

Fig. 1.16 GOES-East satellite view in the infrared of cloud cover over western U.S. and northern Mexico at 7:31 pm Mountain Standard Time on August 28, 2019 [0231 UTC, 29 August] (*courtesy* College of DuPage Meteorology)

Megaflashes that only occur within MCSs are described in Sect. 1.4.5. It is not likely that Arizona can have such megaflashes. However, the monsoon-season MCSs in southern Arizona into Northern Mexico (Figs. 1.15 and 1.16) likely have flashes that last several seconds and extend in cloud for over 100 km. These occasional lightning events are very complex that make the usual definitions of CG and IC difficult to categorize. Nevertheless, the vast majority of ICs in Arizona travel a short distance, last only a portion of a second, and are close in space and time to CG flashes.

1.5.4   Inverted Troughs

During the monsoon season, upper level systems move from the east or south-east toward the west or northwest across Arizona. These are called inverted troughs, easterly waves, or tropical waves; various members of the meteorological community have specific preferences. These inverted troughs during July and August may move steadily from the Gulf of Mexico to reach Arizona at any time of day or night and take a day or two to cross the region. They can be difficult to locate with satellite, radar, lightning, or upper air data, but the possible subtle indications are tracked as much as possible by forecasters.

If the leading edge, where the most lift occurs, reaches the state when the sky is clear and daytime heating is strong, this subtle feature can produce a major lightning and rainfall event with seemingly no warning. Sometimes they stall, sometimes they evolve overhead, and sometimes they dissipate! Inverted troughs are common throughout the tropical and subtropical latitudes around the globe, and some are the originating disturbances for tropical storms and hurricanes to form over the tropical oceans.

1.5.5 Hurricanes and Tropical Storms

As much as half of the rainfall in the southwestern deserts of Arizona results from decaying tropical systems (Stevenson et al. 2016). The outer rainbands tend to have more lightning than the eyewall, and tropical storms tend to have more lightning than hurricane-force systems. In Arizona, lightning is not often directly related to the storm itself but associated with the abundant low-level moisture that is brought from the south or southwest from the Gulf of California (Sea of Cortez) and Eastern Pacific Ocean. Lightning in tropical cyclones in the Eastern Pacific Ocean has been considered for anticipating growth stages by Leary and Ritchie (2009).

1.5.6 Large-Scale Systems

In Arizona during the winter months, cold fronts, low-pressure troughs, and upper lows sometimes produce widespread but infrequent lightning such as an example shown for other parts of the country by van den Broeke et al. (2005).

1.5.7 Rainfall

Lightning frequency is not always directly related to rainfall amount in Arizona. Very large rates of lightning often indicate heavy rain at the ground, but the reverse is not necessarily true (Minjarez-Sosa et al. 2012, 2017, 2019). A threat for forest fire ignition in Arizona is dry lightning, which results from high-based thunderstorms that form above dry lower levels of the atmosphere that are not accompanied by enough rainfall to extinguish a forest fire (Nauslar et al. 2013).

Question

What types of meteorological conditions do you think cause the most lightning in your area? What time of year is the most frequent? What time of day?

References

Bitzer PM (2016) Global distribution and properties of continuing current in lightning. J Geophys Res 122:1033–1041

Cummins KL (2012) Analysis of multiple ground contacts in cloud-to-ground flashes using LLS data: the impact of complex terrain. In: Preprints of the 22nd international lightning detection conference, Vaisala, Broomfield, Colorado, 02–03 April 2012

Dotzek N, Rabin RM, Carey LD et al (2005) Lightning activity related to satellite and radar observations of a mesoscale convective system over Texas on 7–8 April 2002. Atmos Res 76:127–166

Elsom DM (2015) Lightning: nature and culture. Reaktion Books, London, U.K., p 240

Ely BL, Orville RE, Carey LD et al (2008) Evolution of the total lightning structure of a leading-line, trailing-stratiform mesoscale convective system over Houston, Texas. J Geophys Res 113:D08114. https://doi.org/10.1029/2007JD008445

Evans WH, Walker RL (1963) High-speed photographs of lightning at close range. J Geophys Res 68:1265–1375

Fairman SI, Bitzer PM (2022) The detection of continuing current in lightning using the Geostationary Lightning Mapper. J Geophys Res 127:e2020JD033451

Gasper GEM, Tanner BK (2022) A marvellous sign and a fiery globe: a medieval English report of ball lightning. Weather 77:232–234

Hodanish S, Holle RL, Lindsey DT (2004) A small updraft producing a fatal lightning flash. Weather Forecast 19:627–632

Hodanish S, Wolyn P, Mozley K (2015) Meteorological analysis of the Rocky Mountain National Park lightning fatalities of 11 and 12 July, 2014. In: Preprints of the 7th conference on the meteorological applications of lightning data, Phoenix, Arizona, American Meteorological Society, 04–08 January 2015

Holle RL, Murphy MJ (2015) Lightning in the North American monsoon: an exploratory climatology. Mon Weather Rev 143:1970–1977. https://doi.org/10.1175/MWR-D-14-00363.1

Holle RL, Zhang D (2017) So you think you know lightning: a collection of electrifying fast facts. Vaisala, Inc., 64 pp. www.vaisala.com/en/lp/so-you-think-you-know-lightning

Kahl JDW (1998) National Audubon society first field guide: weather. Scholastic Inc., New York, p 159

Laing AG, Fritsch JM (1997) The global population of mesoscale convective complexes. Q J R Meteor Soc 123:389–405

Lang TJ, Pédeboy S, Rison W et al (2017) WMO world record lightning extremes: longest reported flash distance and longest reported flash duration. Bull Am Meteor Soc 98:1153–1168

Leary LA, Ritchie EA (2009) Lightning flash rates as an indicator of tropical cyclone genesis in the Eastern North Pacific. Mon Weather Rev 137:3456–3470

Ludlum DM (2001) Weather. HarperCollins Publishers, London, p 663

Ludlum DM, Holle RL, Keen RA (1995) The Audubon society pocket guide: clouds and storms. Alfred A. Knopf, 192 pp

Ludlum DM, Keen RA, Holle RL, (1991) The Audubon society field guide to North American weather. Alfred A. Knopf, 656 pp

Lyons WA (2006) The meteorology of transient luminous events—an introduction and overview, Chap. 1, NATO Advanced Study Institute, NATO Science Series II (Mathematics, Physics and Chemistry). Springer, Fullekrug M, ed., Corte, Corsica, 225:19–56

Lyons WA, Stanley MA. Meyer JD et al (2009) The meteorological and electrical structure of TLE-producing convective storms. In: Betz HD et al (eds) Lightning: principles, instruments and applications. Springer Science+Business Media B.V., pp 389–417. https://doi.org/10.1007/978-1-4020-9079-017

Lyons WA (2022) Ball lightning: tracking the Sasquatch of meteorology. Weatherwise 75:35–41

Lyons WA, Bruning EC, Warner TA et al (2020) Megaflashes: just how long can a lightning discharge get? Bull Am Meteor Soc 101:115–121

McCollum DM, Maddox RA, Howard KW (1995) Case study of a severe mesoscale convective system in central Arizona. Weather Forecast 10:643–665

Minjarez-Sosa C, Castro CL, Cummins KL et al (2012) Toward development of improved QPE in complex terrain using cloud-to-ground lightning data: a case study for the 2005 monsoon in Southern Arizona. J Hydrometeor 13:1855–1873

Minjarez-Sosa C, Castro CL, Waissmann J et al (2017) An improved QPE in complex terrain employing cloud-to-ground lightning occurrences. J Appl Meteor Clim 56:2489–2507

Minjarez-Sosa C, Waissmann J, Castro CL et al (2019) Algorithm for improved QPE over complex terrain using cloud-to-ground lightning occurrences. Atmos 10:10 pp

Nauslar NJ, Kaplan ML, Wallmann J et al (2013) A forecast procedure for dry thunderstorms. J Oper Meteor 1:200–214

Peterson M (2021) Where are the most extraordinary lightning megaflashes in the Americas? Bull Am Meteor Soc 102:E660–E671

Peterson M, Stano G (2021) The hazards posed by mesoscale lightning megaflashes. Earth Interact 25:46–56

Peterson MJ, Lang TJ, Logan T et al (2022) New WMO certified megaflash lightning extremes for flash distance and duration recorded from space. Bull Am Meteor Soc 103:1243–1247

Rakov VA (2016) Fundamentals of lightning. Cambridge University Press, 257 pp

Saunders CPR (1993) A review of thunderstorm electrification processes. J Appl Meteor 32:642–655

Steiger SM, Orville RE, Carey LD (2007) Total lightning signatures of thunderstorm intensity over North Texas. Part II: mesoscale convective systems. Mon Weather Rev 135:3303–3324

Stevenson SN, Corbosiero KL, Abarca SF (2016) Lightning in eastern North Pacific tropical cyclones: a comparison to the North Atlantic. Mon Weather Rev 144:225–239

Stolzenburg M, Rust WD, Marshall TC (1998) Electrical structure in thunderstorm convective regions. Part 3: synthesis. J Geophys Res 103(D12):14,097–14,108

Uman MA (1964) The diameter of lightning. J Geophys Res 69:583–585

Uman MA (2011) Lightning. Dover Publications, 320 pp

Uman MA (1986) All about lightning. Dover Press, 167 pp

van den Broeke MS, Schultz DM, Johns RH et al (2005) Cloud-to-ground lightning production in strongly forced, low-instability convective lines associated with damaging wind. Weather Anal Forecast 20:517–530

2

Arizonans' Fascination and Perspectives About Lightning

Abstract In this chapter, the critical role of the large range of topographic features in Arizona lightning is presented, in particular, the tall and often isolated mountain ranges, as well as sharp escarpments. Arizona-specific cultural views and types of lightning are described, such as the Thunderbird, the insights from symbols painted by a Hopi artist inside the Indian Watchtower at Desert View within the Grand Canyon, and stories told by a White Mountain Apache tribal member. Arizona sports teams with lightning and thunder in their names are then listed. The chapter concludes with an editorial cartoon by a Tucson newspaper artist about Arizonans' fascination with lightning.

2.1 Arizona Topography

Now that we have seen how lightning forms in cumulus clouds, and how to categorize lightning in Chap. 1, we now turn to the specific case of Arizona. We wrote a booklet titled "So you think you know lightning: A collection of electrifying fast facts" (Holle and Zhang 2017) that is available at https://www.vaisala.com/en/system/files/documents/Lightning-Booklet.pdf and https://lightningdev.umd.edu/aert/Safety.html. There, we responded to questions that we have heard in Arizona, including from University of Arizona students, as well as those asked by the public and media within the state and elsewhere. Chapter 1 describes these commons myths and misconceptions in detail.

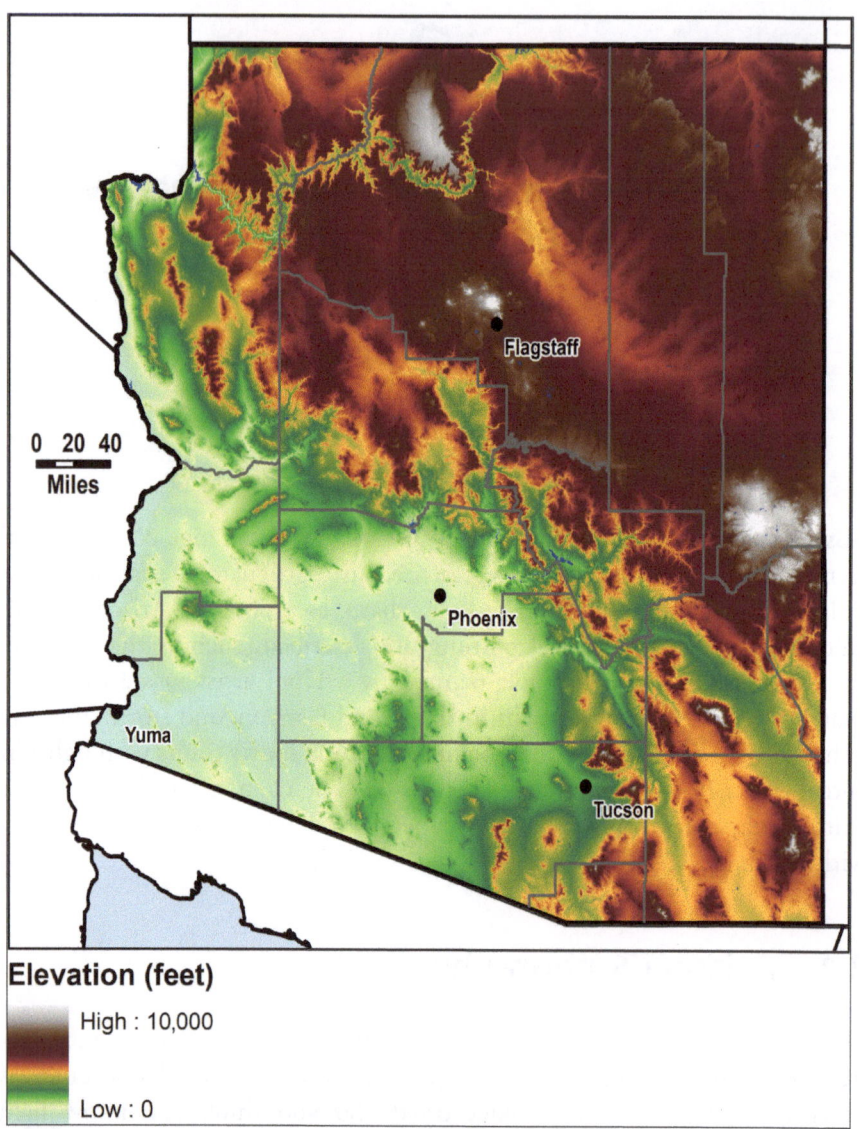

Fig. 2.1 Map of elevation in feet of Arizona with major cities referenced in this book

Let us start with one of the most important ingredients in lightning formation in Arizona—elevation. The altitude varies greatly across Arizona (Fig. 2.1). Yuma in the southwest is at only 43 m (141 ft) along the Colorado River. The highest point is Humphreys Peak at 3,851 m (12,633 ft) north of Flagstaff where a ski resort is located. The higher altitude northern half of the state is separated from the more populous southern regions by the Mogollon

Rim in the central and eastern portions of the state. The White Mountains toward the eastern border with New Mexico also have a high summit on Mount Baldy of 3,475 m (11,400 ft). Numerous isolated mountain ranges in the southeast are sometimes called sky islands since they are isolated as they rise abruptly from the desert. The Chuska Mountains in the northeast near the Four Corners intersection of Arizona with Colorado, New Mexico, and Utah also have a role in thunderstorm formation.

2.2 Lightning Perspectives of Native Americans in Arizona

2.2.1 Overview

Native Americans in Arizona were predominantly rural and engaged in various types of agriculture. This lifestyle engenders watching the weather closely. It has not proven to be easy to find the perceptions of Native Americans regarding lightning, as these may be closely held cultural topics that are oral rather than written.

Arizona has 22 federally recognized tribes as mapped at https://naair.ari zona.edu/sites/default/files/comprehensive_map_reduced_1.pdf comprised of 425,000 enrolled members: about half live on tribal lands. Native American stories relating to lightning specifically for Arizona have been found for only a few of these tribes. Bruchac (1993) describes an elaborate Navajo story about the hero Twins who used "the lightning that strikes straight" and "the lightning that strikes crooked" to kill several mythical beasts that were plaguing The People (Dine) and in the process the Grand Canyon was created. A new publication by Patterson (2021) examines artistic symbols of clouds, rain, and lightning made by Native American tribes in the Four Corners states of Arizona, Colorado, New Mexico, and Utah.

In other areas of the U.S., there is a short collection of lightning and thunder stories by tribes available at http://www.native-languages.org/leg ends-lightning.htm. However, none of them directly include Arizona tribes. Nevertheless, it is very apparent that Native Americans across the U.S. respected, feared, and paid close attention to lightning and thunder. After reviewing numerous books on Native American stories such as Vogel (2001), no additional specific references were found to lightning and thunder from tribal members within Arizona.

Fig. 2.2 The thunderbird in Native American mythology as portrayed on the U.S. Air Force McDonnell Douglas F-4E Phantom II jet at the Pima Air and Space Museum in Tucson, Arizona (©R. Holle)

2.2.2   The Thunderbird

The thunderbird is a common mythological figure in many Native American cultures, including that of the Navajo tribe that occupies a large area of northeastern Arizona. A common thunderbird representation that is seen in Arizona to the present day is shown in Fig. 2.2 and in rock art from neighboring New Mexico at https://www.desertusa.com/desert-activity/mystical-icons.html. The Thunderbird School of Global Management within Arizona State University is located in Phoenix. How did that name result? It is named after a World War II airbase that was located at the present-day site of the school. Further back, the airbase was named after this prominent Native American symbol. The name also appears in the names of streets, offices, and businesses throughout the Phoenix and Tucson areas, and likely other locations in Arizona; there are too many to list! Another common symbol is a simple jagged lightning that is often included in photos of drawings and rock art such as those on display at the Arizona State Museum on the University of Arizona campus. Of course, the thunderbird name also applies to cars and other contemporary items associated with speed and power. In addition, the U.S. Air Force Thunderbird jets regularly perform at air shows, including the "Thunder and Lightning" event over Tucson's Davis-Monthan Air Force Base.

Air Force Thunderbirds

The U.S. Air Force Thunderbird demonstration team has had the Native American thunderbird emblem (Fig. 2.2) as its logo since they were activated in 1953 as the 3600th Air Demonstration Team at Luke Air Force Base west of Phoenix.

2.2.3 White Mountain Apache

A fascinating book by Watt and Basso (2004) was referred to us by Diane Dittemore at the Arizona State Museum on the University of Arizona campus. The book contains the following paraphrased stories up to a century old as told by Eva Tulane Watt of the White Mountain Apache tribe that occupies the east-central portion of Arizona where there is frequent lightning (Chap. 4):

- Mrs. Watt saw one of her brothers die at a boarding school in Rice (Arizona) after handling pieces of wood charred by lightning (pp. xix–xx).
- A medicine man was hit by lightning that resulted in a scar down his legs. He was not supposed to eat for several days but recovered with corn soup and a boiled little red plant (p. 26).
- If you are hit by lightning, a squash or corn plant dries up if it's touched by a person (p. 28).
- Ladies can't have children due to infertility after being hit by lightning (p. 28).
- Lightning goes after anything with a sharp point such as little yuccas, so they are chopped down and thrown away if they grow into larger plants that are then too close to the wickiups or houses (p. 28).
- Lightning strikes pine trees but not sycamores.
- People affected by lightning were given some kind of boiled herbs by Mrs. Watt's grandfather and then given a bath with it. The wounds were washed, and the person drank four dippers of the liquid (p. 28).

- A wickiup hit by lightning was torn down by four boys who inhaled the smoke and died within a month (pp. 56–58).
- A group of boys tending cattle near Ash Creek was killed when they went under a pine tree that was hit by lightning (pp. 62–63).
- A lady from San Carlos was in a wickiup hit by lightning that started burning and produced smoke. She drank a liquid made from the *i'dii ch'il* (thunder bush) and recovered (pp. 63–63).
- Some whitemen recovering salt from the banks of the Salt River were killed at their camp by lightning. The Apache viewed this as justice for desecrating a sacred place (pp. 171–173).
- Her mother went out the back door of a house during a thunderstorm and was hit (pp. 281–282).
- A person hit by lightning is not supposed to be taken across a river. Yellow powder needs to be sprinkled on the river first.

2.2.4   Hopi Artwork at Grand Canyon National Park

Hopi traditions about lightning are included on the inside walls of the Indian Watchtower at Desert View located toward the East Rim of Grand Canyon National Park. Architect Mary Colter designed the Watchtower that was completed in 1933, and Fred Kabotie of the Hopi tribe painted the inside with traditional images common to his tribe located east of the Grand Canyon in northeast Arizona.

A variety of lightning themes is visible in the Circular Painting in the Hopi Room on the first floor of the Watchtower (Fig. 2.3). These descriptions of the images are extracted from a booklet written by Watchtower architect Colter (1933):

- Outside the main circular design, on the four corners, are clouds that have rounded tops and rain represented by straight lines downward from the clouds. In particular, there are long forked symbols that extend outward to the upper right and upper left—these represent lightning, who are messengers of the gods.
- Inside the circle in the upper right panel is a stylized version of these same features: a heap of squares for the cloud, falling rain, and lightning shown by stepped jagged lines.
- The lower left panel shows a row of six rain showers and a man holding a shield dripping rain, but no lightning.
- Further outside the circle in the upper left of Fig. 2.3 is a symbol that is the thunderbird mentioned in Sect. 2.2.2.

Fig. 2.3 Hopi traditions about lightning painted by Fred Kabotie on the inside walls of the Indian Watchtower at Desert View in Grand Canyon National Park. [Bildagentur-online/Schickert/Alamy Stock Photo]

It is no coincidence that the rain ends abruptly below the cloud. Hopi land is on a high plateau (Fig. 2.1) where it is very dry. As a result, virga is common, that is, rain falling from a cloud but not always reaching the ground. However, lightning can and does occur from clouds with only virga.

The second author of this book visited the Carnegie Museum of Natural History in Pittsburgh, Pennsylvania and quite surprisingly found many of the same images in the Kabotie paintings. This exhibit was based on material collected on the Hopi Reservation in 1904. A photo in the exhibit (Fig. 2.4) shows a window in a home on a Hopi village with lightning, cloud, and rain symbols nearly identical to shapes in the four corners of Fig. 2.3—but made of stone! Other parts of the exhibit include rain cloud images very similar to those in Fig. 2.3. Further, a wavy line representing lightning was made into a wand carried by males in the Buffalo Dance. All of these shapes and objects closely match those painted by Kabotie.

A classic book on Arizona climate was published by the University of Arizona Institute of Atmospheric Physics in cooperation with the U.S. Weather Bureau (now the National Weather Service) by Green and Sellers (1964). Although it has no lightning or thunderstorm data, it is a fascinating detailed look at the monthly temperature and rainfall of every known weather station in Arizona, sometimes back to the 1890s. Remarkably, its

Lynn Dalton under rain cloud stone relief near Hotevilla, Arizona. *(Photo by Owen Seumptewa, ca. 1978)*

Fig. 2.4 Photo of Hopi Reservation window with lightning, cloud, and rain symbols as exhibited at Carnegie Museum of Natural History in Pittsburgh (© D. Zhang)

cover includes the thunderstorm and lightning symbols shown at the four outer corners of the Kabotie painting in Fig. 2.3. Sellers and Hill (1974) updated a number of the entries in the earlier edition. If you want to study the climate of Arizona at obscure as well as better known places in Arizona, these books are for you!

The booklet by Colter (1933) also has several other references to lightning and thunder at the Watchtower, indicating how important and noteworthy such an event appears to Native Americans, the Hopi in particular:

- Bullroarer: One image shows a figure holding a bullroarer, a stick that when whirled quickly makes the sound of distant thunder.

- Double lightning: A pair of jagged shapes represents lightning in a form that is stated to be common in Native American sites identified in and around the Grand Canyon.
- Lightning frames: One of the Kabotie paintings has figures with lightning framers or "whizzers" that typify lightning and are shot toward the end of Hopi ceremonies.
- Shield: A warrior has a shield design with lightning rods projected on four sides of it. The lightning rods are said to be the "Powers of Destruction."
- Thunderbird: This symbol (Fig. 2.2) was used as a trademark for 25 years in various forms by the Fred Harvey Company who managed concessions inside Grand Canyon National Park. It comes in many shapes and may become somewhat abstract.

A recent study of prehistoric art by Patterson (2021) shows cloud and weather imagery from 1000 BC to 1300 AD in the Four Corners region. The images include not only lightning and thunder, but rain clouds, rainbows, hail, and virga in generally similar shapes as in the Kabotie drawings at the Watchtower in Fig. 2.3. This online publication does not isolate the symbols to Arizona, but the commonality of lightning and thunder as an important part of life myths hearkens back several thousand years in this region! Not only is lightning perceived as important, but it also is associated with life-giving rain that is so critical to agricultural dwellers of this arid region.

Future Research

Aside from these resources, not much is written about Native American lightning stories. It is an open topic for future research to collect ideas from the 22 Arizona tribes. However, such legends or stories are at the core of their history and may not be readily revealed to non-Native Americans.

2.3 Arizona Sports Teams with Lightning and Thunder

Weather is a common source of nicknames for sports teams at all levels, including heat, wind, tornadoes, monsoon, dust devils, and nearly every meteorological phenomenon. Lightning and thunderstorms are often used in team names since they refer to such qualities as strength, speed, and danger.

Outside of Arizona, the Tampa Bay Lightning of the National Hockey League is named for the region's high ranking in the frequency of lightning in the U.S. The Oklahoma City Thunder of the National Basketball Association refers to the frequent severe storms that occur in that area.

For Arizona, a starting point is the article that appeared in the *Arizona Daily Star* newspaper in Tucson on March 03, 2019. Teams in the Tucson area with thunder references include the following:

- Arizona Thunder Blades: An inline hockey team in 1996 that folded after three games.
- Old Pueblo Lightning: A currently active amateur women's rugby team in Tucson that was not included in the 2019 newspaper article.
- Sonoran Thunder: A women's soccer team that played in 2002–2003.
- Tucson Thunder Kats: An indoor football team that attempted a 2011 start but never played.

The pervasiveness of thunder and lightning is so broad that it is not possible to identify all of the teams at various levels with lightning and thunder in their name. One notable example is the girls' competitive fastpitch softball team called the Arizona Desert Thunder. It is currently coached by Jorge Campos, who fittingly is an employee of Vaisala in Tucson and is part of a group that establishes and maintains antennas for lightning detection around the U.S. and world. Some other Arizona girls' softball teams have the names Thundercats, Thunder, and Storm.

At the high school level in Arizona, here are schools with names related to lightning, thunder, and the monsoon:

- Desert Vista High School Thunder in Phoenix.
- Lincoln Preparatory Academy Lightning in Chandler.
- Mica Mountain Thunderbolts in Vail.
- Mohave High School Thunderbirds in Bullhead City.
- Thunderbird High School Chiefs in Phoenix.
- Valley Vista High School Monsoon in Surprise.

Other Teams

How many weather phenomena are the source of sports teams in your area? How many have a variation on lightning and thunder? It would be interesting to collect them to compare with the most common type of weather where you live!

2.4   Arizona's Fascination with Lightning

Arizona residents' fascination with lightning is illustrated by the 2009 cartoon in Fig. 2.5. David Fitzsimmons is an editorial cartoonist and columnist for the *Arizona Daily Star* newspaper in Tucson. Over the years, he often features monsoon-related cartoons, and his comment about this figure is that "You can't reside here and not love lightning." In terms of the thunderstorm, he has both cloud-to-ground and intra-cloud lightning (Sect. 1.4), thunder, and a microburst. Other features include factors explored in Chap. 3 of saguaro cactus and distant mountains, along with other iconic Arizona images of a quail, jackrabbit, tumbleweed, and javelina!

Fig. 2.5 Cartoon by David Fitzsimmons of the *Arizona Daily Star* (by permission from David Fitzsimmons)

References

Bruchac J (1993) How the hero twins found their father, flying with the eagle, racing with the great bear. BridgeWater Books, Mahwah, NJ, p 128

Colter MEJ (1933) Manual for drivers and guides: descriptive of the Indian Watchtower at Desert View and its relation, architecturally, to the prehistoric ruins of the Southwest, Fred Harvey Company. Reprinted in 2017 with added photos by the Grand Canyon Association, Grand Canyon, Arizona, p 104

Green CR, Sellers WD (1964) Arizona climate. University of Arizona Press, Tucson, p 503. ISBN-13: 978-0816502769

Holle RL, Zhang D (2017) So you think you know lightning: a collection of electrifying fast facts. Vaisala, Inc., p 64. https://www.vaisala.com/en/system/files/documents/Lightning-Booklet.pdf; https://lightningdev.umd.edu/aert/Safety.html

Patterson C (2021) Clouds in the prehistoric art of the Colorado Plateau. Expression, Atelier editions in cooperation with UISPP—CISENP, no 33, pp 43–64

Sellers WD, Hill RH (1974) Arizona climate, 1931–1972, 2nd edn. University of Arizona Press, Tucson, p 616. ISBN-13: 978-0816504664

Vogel CG (2001) Weather legends: native American lore and the science of weather. Millbrook Press, Brookfield, Connecticut, p 80

Watt ET, Basso KH (2004) Don't let the sun step over you: a white mountain apache family life, 1860–1975. University of Arizona Press, Tucson, p 340

3

Arizona is the Lightning Photography Capital of the U.S

Abstract Pictures of lightning in Arizona are widely known as among the best anywhere in the U.S. and perhaps the world. The combination of large mountains on the horizon in the background, frequent colorful sunsets, large saguaro cactus, and high cloud bases that allow views of tall lightning channels make for an attractive unique view. For these reasons, more aesthetic lightning photos are likely taken in Arizona than anywhere in the U.S. Examples from current Arizona photographers are shown in this chapter, along with methods and tips of how to photograph lightning. Special attention is given to the first known Arizona lightning photo from the early 1910s taken at the Grand Canyon.

3.1 Landscape, Geographic, and Meteorological Factors

Arizona is not the lightning capital of the world. Arizona is not the lightning capital of the United States. But it is easily the lightning photography capital of the U.S. in terms of a single location. Figure 3.1 shows the U.S. map of cloud-to-ground (CG) flashes from 2011 to 2020 where it is apparent that Arizona does not have nearly as much lightning as many other locations. Nevertheless, there is considerable variation across the state that will be reviewed in detail in Chap. 4. How pervasive are the misconceptions? For example, the sign shown in Fig. 3.2 along a trail in Saguaro National Park East near Tucson declares that "Tucson is the 'lightning capital' of the U.S."

© Springer Nature Switzerland AG 2023
R. L. Holle and D. Zhang, *Flashes of Brilliance*,
https://doi.org/10.1007/978-3-031-19879-3_3

It appears that the sign has some age, so hopefully we have come a long way since this was printed!

In addition to lightning photography, Arizona has another claim to fame in lightning, since it can be considered the birthplace of modern lightning detection not only for the U.S. but the world. Chap. 6 describes the development of detection methods at the University of Arizona. Chap. 7 relates how some of this development was based on earlier photographic studies at the university from the viewpoint of basic research. It was this foundation that brought together faculty and staff to develop lightning detection such that it has now spread worldwide in many forms.

But first, we have located what is among the first Arizona lightning photographs! Fig. 3.3 shows lightning over the Grand Canyon taken from the South Rim. The Kolb brothers set up a photography shop on the South Rim and took photos of anything and everyone in and around the canyon. Part of their income was from taking photos of visitors who rode mules to the bottom of the canyon—after they paid a toll to the brothers for use of the trail, since it was not yet a National Park! The Phoenix broadcast meteorologist Royal Norman pointed out to us that Northern Arizona University had obtained the Kolb brothers collection of photos and noticed that lightning

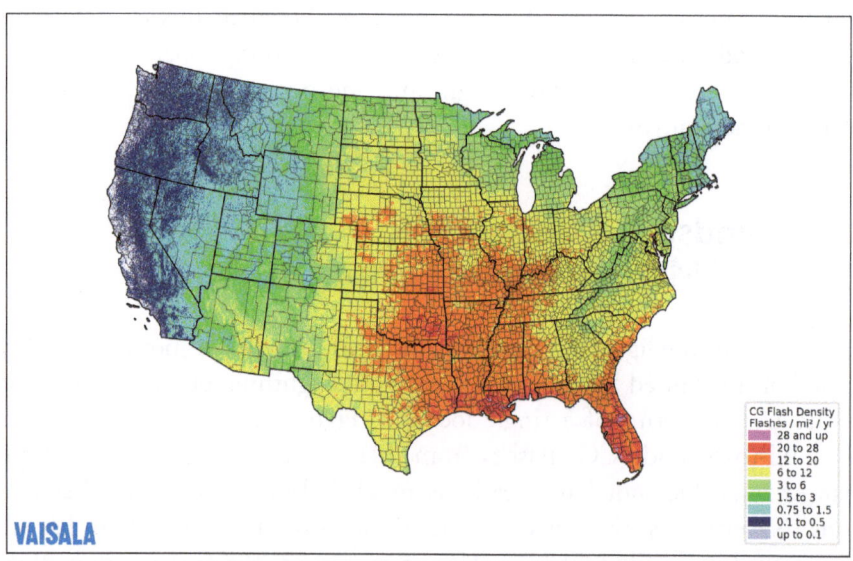

Fig. 3.1 Density of CG flashes per square mile per year over the U.S. on a grid of 3.219 by 3.219 km (two miles by two miles) based on an average of over 21 million CG flashes per year from Vaisala's National Lightning Detection Network (NLDN) from 2011 through 2020. Scale in lower right. Updated from Holle et al. (2021). © American Meteorological Society. Used with permission

Fig. 3.2 Trailside sign in Saguaro National Park East (© R. Holle)

photos were included. The view in Fig. 3.3 was collected on a glass plate that was used early in their photography business. The plates are not dated, but the Northern Arizona staff have been very helpful in identifying this to be one of several lightning pictures taken between 1910 and a few years later. We can see the familiar downward branching of CG flashes as described in Chap. 1. Also, it is very likely that this photo was taken over some period of time, perhaps on the order of a few minutes, to obtain so many flashes in the field of view.

Photos by the Kolb brothers were famous and influential

Their photos were part of the impetus for the formation of Grand Canyon National Park in 1919. It's amazing to see that the *National Geographic* magazine issue of August 1914 features a whopping 85 pages of photos and narratives by the Kolb brothers about their trip down the Colorado River and other adventures in the canyon area!

Take a look online and at calendars, magazines, and other representations of attractive lightning. Often you will see mountains on the horizon, colorful sunsets, saguaro cactus, and high cloud bases with tall lightning channels. In fact, the first author had a collection of Arizona lightning post cards in his office cubicle for many years (Fig. 3.4). These are only from Arizona; probably no other area of similar size in the world has such a concentration of post cards with lightning on them! They have been collected in the last decade

Fig. 3.3 Lightning over the Grand Canyon photographed by Emory Clifford Kolb from the South Rim in the early 1910s on a glass plate. Image 119,947 from colorado plateau archives, special collections and archives, cline library, Northern Arizona University

at drug stores, gift shops, national parks, visitor centers, and other places where cards are sold; there are likely to be more. Beware that unfortunately a large percentage of lightning photos that are seen online and sometimes in publications are deliberately fabricated and are fake.

Let's examine the variety of features that make Arizona such a special place for lightning photos:

- **Mountain ranges on horizon:** Figure 2.1 shows the terrain over the state to be highly variable in most regions. Figure 3.5 shows a view that is not unusual of multiple layers of mountains in the distance.
- **Colorful sunsets:** Due to the dry, clear, and clean air during much of the year, the sun is shining all the way down to the horizon such that sunsets are an event to be checked every evening, and colors and shapes abound (Fig. 3.6).

Fig. 3.4 Post cards of Arizona lightning (© R. Holle)

- **Saguaro cactus:** The symbol of southern Arizona is the saguaro cactus that can have fantastic shapes (Fig. 3.7). They are widespread below about 1,220 m (4,000 ft) throughout the area, and Saguaro National Park is dedicated to them on the east and west sides of Tucson. Saguaros are the

Fig. 3.5 Arizona mountain ranges on the horizon at sunset. This view from the Arizona-Sonora Desert Museum west of Tucson looks toward the Quinlan Mountains on the Tohono O'odham Nation. The rectangular-shaped National Solar Observatory telescope of the Kitt Peak National Observatory is visible to right of center on the ridge line (© S. Holle)

subject of a recent book from the University of Arizona Desert Laboratory on Tumamoc Hill by Bird (2021) that is available online. Maps of their range such as at https://www.nps.gov/sagu/learn/nature/location.htm show the saguaro to occur in the southern and southwestern regions of Arizona into northern Mexico.

- **High cloud bases/tall lightning channels**: Especially during the Southwest Monsoon months of July through September, the lower levels of the atmosphere are often dry, although higher levels have enough moisture to produce thunderstorms. Lightning in Arizona during the monsoon season originates at high altitudes, but then propagates toward the ground (Chap. 1.4) through air that may have only very light rain. Figure 3.8 shows clouds over the San Francisco Peaks north of Flagstaff where the highest peak is at 3,851 m (12,633 ft). Such high cloud bases allow seeing tall lightning channels from a long distance away. As a result, lightning channels that are obscured within clouds in more humid locations are often exposed in Arizona as they travel from their origin in the sub-freezing layer in the cloud to reach the ground (Fig. 1.9). In fact, the cloud bases may

Fig. 3.6 Colorful sunsets viewed from the University of Arizona campus (**a**), Oro valley (**b**), and the Arizona-Sonora Desert Museum (**c, d**) near Tucson in Arizona (Photo a © D. Zhang; photos b, c, d © R. Holle)

be so high that they occasionally are near the typical summertime freezing level between 4.2 and 5.2 km (14,000 and 17,000 feet) in Arizona.

3.2   Examples from Arizona Lightning Photographers

3.2.1   Challenges in Photographing Lightning

Given the common ingredients of mountains on the horizon, sunset colors, saguaro cactus, and tall lightning channels, it is not surprising that photographers who specialize in lightning photography have emerged in Arizona, especially at the lower elevations of the southern counties.

First, let's explore how to photograph lightning. The most important virtue is to be prepared and be patient! Here is why:

- The storm is in a great location but you're at the office or a restaurant and do not have a camera ready.

Fig. 3.7 Saguaro cactus at the base of the Santa Catalina mountains in Oro Valley, Arizona (© S. Holle)

- The storm seems to be over; the camera gear has been packed away, then another lightning event occurs in the perfect place where you had been focusing.
- The storm is raging outside your window so it's too close to use the camera safely.
- There are too many power poles and structures to make an attractive photo.
- A storm develops but by the time you reach it, the lightning is done.
- A car drives into the field of view at night and its light saturates the photo.
- An aircraft flies through the view at night and leaves a blinking red light.
- A storm is in the distance, but you can't find a clear view.
- The wind is too strong to hold the camera steady.
- It starts to rain on the camera.
- The film runs out (more of an issue in prior years).
- The battery dies.

Fig. 3.8 Cumulus congestus with high cloud bases above San Francisco Peaks near Flagstaff, Arizona (© R. Holle)

Fig. 3.9 Lightning around Kitt Peak National Observatory on June 04, 1972 (© Gary Ladd)

To avoid these impediments, be patient and be prepared! Have the camera equipment ready if there is any possibility of a thunderstorm. Stake out some pre-set locations that look in specific directions if and when a storm is forming there and wait.

3.2.2 Arizona Lightning Photographers

Any list of lightning photographers in Arizona will be incomplete or inaccurate, and we apologize for perspectives that may not be correct. The fantastic Arizona scenery continues to attract more and more photographers, so we don't know all of them and by the time the book is published, there will be more lightning photography specialists than before! The following list is mostly chronological in order of the longevity or influence in developing the art form that is shown by their portfolios of lightning pictures taken in Arizona, and includes some lightning photographers that have had widespread influence in various ways:

- **Leon Salanave:** Let's start with the classic lightning book that has been an inspiration for us. The book by Dr. Leon Salanave (1980) is notable for several reasons:

  1. It is based at the University of Arizona.
  2. Many of the photos are from Tucson.
  3. It presents breakthroughs in lightning science at the University of Arizona as covered more fully in Chap. 7.
  4. Some of the lightning photos are published in color.
  5. It was published in 1980 during the transition from basic lightning research to the development of modern lightning detection technology at the University of Arizona.
  6. The scientific content was up to date as of 1980 but it can also be appreciated for its collection of lightning photos that continue to be remarkable.
  7. Some of his photos appear in the classic book, *All about lightning* by Uman (1986).

Salanave (1980)

This book is available online at a reasonable price and is one to have for an occasional read. The first author bought several of the last copies that the University of Arizona bookstore had in stock in the early 2000s.

- **Gary Ladd**: The photo by Gary Ladd in Fig. 3.9 taken at Kitt Peak is widely known and published and helped launch his career in nature and landscape photography. Note that Kitt Peak is also shown in Fig. 3.5 from afar. Ladd also collaborated with Dr. E. Philip Krider at the University of Arizona on analysis of upward streamers (leaders) from a CG flash photographed from Kitt Peak (Krider and Ladd 1975). Since taking this iconic photo while studying astronomy at Kitt Peak, his mainly Northern Arizona photos have appeared in *Arizona Highways, National Geographic, Smithsonian,* and other publications about the southwest U.S. He is based in Page in northeast Arizona.
- **Ralph Wetmore**: Ralph Wetmore personifies the dedicated lightning photographer in Tucson. He has probably taken more lightning photos in southern Arizona, over a longer time period, and spent more personal time exploring this topic than anyone. Wetmore (2010) is a self-published book with an amazing array of color lightning photos from the Tucson area. His favorite approach is to go to one of his pre-selected locations around the city, such as parking garage roofs, and wait. Originally, he was the only person at a spot when he arrived, but now he is rarely the only one to pick that place and time. Our online booklet (Holle and Zhang 2017) includes two of his photos. It should be pointed out that he is not making a living from lightning photography alone, as is the case for all of the following entries—it is a passion!
- **Tom Wiewandt**: Tom Wiewandt is the author of a variety of publications about nature, environment, and other topics in the southwest, and is based in Tucson. The remarkable nature photography book by Wiewandt and Wilks (2001) about the southwestern states has two lightning photos from Arizona. He also photographs animals, plants, geologic formations, and other features of the southwest. From this material, he has developed widely acclaimed videos and other digital and print resources that include lightning. His video complement to the book is frequently shown for fund-raising on Tucson public television.
- **Warren Faidley**: Warren Faidley is also one of the original lightning photographers in the Tucson area. His first widely circulated lightning

photo showed a CG flash hitting a light pole in a Tucson oil and gasoline tank farm that was published in *Life* magazine in 1989. His photos are inside the August 1990 issue of *Arizona Highways*, and his photo of lightning from an isolated thunderstorm over Tucson is one of the iconic such photos of the last few decades that filled a two-page spread in the July 1993 *National Geographic* article on lightning. The book by Kramer and Faidley (1992) contains Tucson-area lightning photos, including the famous tank farm picture as well as sunset photos with lightning. Lightning photos by Faidley are on the cover of the *Audubon Society Pocket Guide* (Ludlum et al. 1995) and *Audubon Society First Field Guide* (Kahl 1998), in the British book *Weather* by Ludlum (2001), and on the cover and inside Parzybok (2005). One or more of these books are often available in bookstores at national parks and similar locations. His tank farm and other lightning photos are also within the full color illustrated book for Grades 4–6 by Harris (2005).

Book editors

Based on Faidley's entries in weather-related books and many other outlets not listed here, it is obvious that book editors love Arizona lightning photos to enliven their publications! Needless to say, many additional Arizona photographers who do not specialize in lightning also take photos of this phenomenon.

- **Mike Olbinski:** Michael Olbinski is based in Phoenix and photographs lightning and other types of weather in Arizona. He is among the first to obtain time lapse sequences of haboobs, the desert sandstorms forming an advancing wall that overtake the Phoenix area during the monsoon season. Other Olbinski photos of Arizona phenomena include microbursts, dust devils, and thunderstorms during the monsoon months. An example of his lightning photos is shown in Fig. 3.10. Another remarkable photo by Olbinski taken in New Mexico a short distance east of southeast Arizona (Cummins et al. 2018) shows upward leaders surrounding two CG strokes in the open desert. These upward leaders emanated from small grass and shrubs; the relevance of this information is described in the lightning injury mechanism list in Chap. 5.2.
- **Greg McGown:** Greg McGown is a relatively new weather photographer in the Tucson area. His photo shown in Fig. 3.11, and on the cover of this book, is an amazing juxtaposition of the ingredients listed above, with the

Fig. 3.10 CG flash on September 13, 2017 on the north rim of the Grand Canyon emanating from a high cloud base at sunset in Arizona (© Mike Olbinski)

addition of a rainbow. He is a Realtor and not currently making a living from photography.

- **Lori Grace Bailey**: Lori Bailey is also a recent addition to the weather photography scene in southern Arizona. She lives near the Arizona border with Mexico and photographs lightning with different foregrounds such as the view shown in Fig. 3.12 that are different than the more common Tucson views. Her lightning photography was the focus of a recent article in the monthly Tucson magazine Desert Leaf (https://online.fliphtml5. com/uyvk/ncbe/#p=1).

- **David O. Rankin**: David Rankin resides in southern Utah and travels to northern Arizona to photograph all types of natural phenomena. The view of Arizona lightning shown in Fig. 3.13 was taken in the area of Page next to Lake Powell.

Flash flood video

Check out the amazing videos of dangerous flash floods resulting from debris flow into normally dry washes and canyons on Rankin's website at http://ran kinstudio.com/flashfloods!

Fig. 3.11 CG flash from high cloud base with a saguaro, distant mountains, and a rainbow in the late afternoon of August 08, 2015 near Tucson, Arizona (© Greg McGown)

Fig. 3.12 CG flash from a high cloud base with distant mountains and sunset on August 27, 2019 near Rio Rico in southern Arizona (© Lori Grace Bailey)

Fig. 3.13 CG flashes from high cloud bases near Lake Powell. Page, Arizona is on the right and the now dismantled smokestacks of the Navajo Generating Station are on the left (© David O. Rankin)

- **Additional Photographers**: At the risk of slighting anyone, we can't show photos for all of the well-known lightning photographers in Arizona. The above are people that we have personally encountered through various organizations. One of them is the Southeast Arizona chapter of the American Meteorological Society (https://www.ametsoc.org/chapters/seacams/) where we have been officers and invited several of these photographers to speak.

The following are additional Arizona-based photographers who include high-quality lightning views in their portfolios and sometimes have photos accessible on the web:

Flagstaff.
–Dr. David Blanchard.
–Jeremy Perez.
Tucson.
–Josh Adams.
–Robert Campbell (samples in Wetmore 2010).
–John Sirlin.
–Bryan Snider.
–Val Valdez.

–John Walden (samples in Wetmore, 2010).

–William Wantland (Tucson lightning photos on cover of August 1990 issue of *Arizona Highways* magazine and in Brunet et al. 1999).

–Tony Winebarger (samples in Wetmore 2010).

–Kent Wood (samples in Wetmore 2010).

Southeast Arizona.

–Glenn Weeks.

Newspaper three-page spread on lightning

Tucson is such a prime lightning photography location that the *Arizona Daily Star* newspaper in September 2008 devoted three color pages to the topic from the Tucson area. None of the chosen photos came from the photographers shown earlier in this chapter, indicating the widespread fascination with lightning photography in this area.

3.3   Lightning Photography

Safety is always the foremost important factor! Photographing from inside a car or substantial building for safety is highly recommended. Needless to say, using a tripod is necessary. A cable remote can be very helpful. The second author has taken lightning videos from the penthouse on the roof of the University of Arizona Atmospheric Sciences department. So, pre-selecting a few safe places such as what some of those professional photographers do is also important!

3.3.1   Film

Most of the above photographers started taking lightning pictures with film cameras. Many of the principles learned with film are worth recounting since they apply to digital cameras. First and foremost, they learned to be selective and careful in exposing the film. Film costs money, and processing costs money. Then there's the issue that when the film comes back from being processed, you find out what worked and what did not. A general rule is that only one of ten frames that are exposed have lightning on them, and many of those are not especially interesting. So, the photographer with film learns to pay attention to the storm, try to guess the interval between visible lightning events in the field of view, estimate the correct shutter speed and f-stop, and hope for the best.

The approach is to expose the film for as long as it takes for a lightning event to occur within the field of view or think it did! The length of exposure is measured in seconds, or until the field of view has too much other light filling the frame, then that exposure is closed. The B (for bulb) setting is used to keep the shutter open. Do you open the lens to allow the full amount of light onto the film—no! It is an arbitrary decision about what to use for a setting, but around f5.6 is a starting point. The desired effect is to show the lightning channel as well as other features such as sunset or city lights but not so bright that the lightning blasts the entire photo to make it white. So, making a good lightning picture is a matter of experience by taking into account the distance to the lightning and how much ambient light is in the field of view from other sources. In addition, what other man-made or geographical features do you want to include? Also, you can control to some extent whether the photo should have only one lightning event or an accumulation of them over several seconds or minutes. Considering all of these often-uncontrollable variables, it is gratifying to take a photo that shows lightning clear enough to be interesting and/or attractive!

The first author took weather and cloud photos for decades using Kodachrome 64 film due to its bright colors, fine grain, and excellent storage life. This film had an ISO of 64. But Kodachrome 64 was discontinued in 2009. Film from Fuji and other manufacturers were also popular for cloud and weather photos when Kodachrome 64 was no longer available.

Scanning of film photos is an extra step to make prints from slides or negatives. Every few years, a new scanning capability is on the market and a learning curve can be required to use the software and hardware, depending on the desired quality. Since there are so many Kodachrome photos of lightning, some of the best pictures have been scanned over the years with great success. This is done because some views are so difficult to obtain that they are worth archiving! It is also possible to introduce a certain amount of subjectivity into the scanned photo, such as making segments lighter or darker.

Scanning

Scanning is not unlike the process of "dodging" prints made from film negatives to obtain a desired effect. The iconic photographer Ansel Adams was particularly adept at dodging his prints to make them into artwork. Some of his prints are at the University of Arizona Center for Creative Photography, but he does not seem to have taken any lightning pictures.

3.3.2 Digital

The digital camera has one large advantage—not buying lots of film, paying for processing, then waiting for days to find that most photos don't turn out to be worth keeping. With digital, the results can be seen in real time, although the details may not be apparent until they are shown on a larger screen than the viewfinder on the camera. Nevertheless, most of the underlying concepts are the same as with film…patience, choices of subject view, frustration, uncertainty, and so on. Many of the detailed features involved in lightning photography are specific to each make and model of camera.

There is a wide variety of recommendations and stories about lightning photography with digital cameras. Two of the many useful sites are from Spokane (https://www.spokesman.com/stories/2019/jul/25/lightning-in-the-lens-tips-and-tricks-for-photogra/) and Saskatchewan (https://www.cbc.ca/news/canada/saskatchewan/photography-tips-lightning-1.5235827). Note that lightning safety advice here and at other sites is often not accurate. Instead, go to the National Lightning Safety Council (http://lightningsafety council.org/); both authors of this book are members of the Council, and also see Chap. 5.2.

Some comments on these sites include the following:

- "You'll end up taking a bunch of photos and only a handful will work out, but that's the nature of lightning photography."
- "You never know when a lightning bolt will strike, and so the longer the exposure, the more likely you are to catch the lightning."
- "Thirty seconds has always worked well for me. The real key is get ahead of the storm…"
- "It is a luck of the draw type thing what you capture."
- "…taking photos of lightning is tricky."
- "I took about 300 images and walked away with four that I was happy with."
- "…the photographic variables are difficult to overcome, and the conditions don't usually make for easy shooting."
- "…the actual process of shooting lightning…can be quite adrenalin-filled."
- "It takes a lot of practice to get a good shot of lightning. Be patient. Expect that most of your shots won't work out."

Conclusion

What's the conclusion from this litany of problems? Lightning photography is not easy but it's rewarding! If you want to take on a challenge, try it (safely)!

3.3.3   High Speed

Over the last two decades, the understanding of lightning has been transformed in part by the development of high-speed cameras. At first, tens of frames per second were taken of lightning, then hundreds, then thousands, and now tens of thousands of frames per second. The resolution is now at such a time scale that new phenomena are being observed that were not known only a decade ago (Montanyà et al. 2012; Saba et al. 2009, 2010; Schumann et al. 2022; van der Velde et al. 2019).

Samples of these amazing videos can be viewed by using an online search engine with words such as "warner high speed lightning videos." Although not based in Arizona, a leading photographer is Tom Warner (https://ztrese arch.blog/aboutztresearch/). He has been at the forefront of high-speed video of lightning for more than a decade and has provided the lightning community with new phenomena that have not been seen before with each advance in frame speed!

3.3.4   Lightning Trigger

Devices began to be sold about 20 years ago called lightning triggers. They are designed to link with digital cameras and only expose the lens to lightning when it is occurring, so that not so many unusable images are collected. The problem is that many or most triggers operate after the first stroke. As shown in Fig. 1.11 in Chap. 1.4.2, the first stroke of a negative CG flash has downward branching, but subsequent strokes are mostly not branched. The result is that lightning triggers tend to show an unbranched pencil-thin channel. The underlying issue is that lightning is so fast that it is potentially difficult to link the trigger with the ability of the camera to open the exposure in time. Some triggers may be able to overcome this capability with certain cameras. It should be mentioned that the trigger allows more daytime lightning photography, albeit of single unbranched channels in many or most photos.

References

Bird WL Jr (2021) In the arms of saguaros: Iconography of the giant cactus. In: Proceedings of the Desert Laboratory, Desert Laboratory on Tumamoc Hill, Contribution No. 3, University of Arizona, Tucson, pp 236

Brunet R, Holle R, Mogil HM et al (1999) Discovery channel weather: An explore your world handbook. Discovery Books/Random House, New York, p 192

Cummins KL, Krider EP, Olbinski M et al (2018) A case study of lightning attachment to flat ground showing multiple unconnected upward leaders. Atmos Res 202:169–174

Harris C (2005) Wild weather. Kingfisher Voyages, Kingfisher, Boston, pp 60

Holle RL, Zhang D (2017) So you think you know lightning: A collection of electrifying fast facts. Vaisala, Inc., pp 64 (https://www.vaisala.com/en/system/files/documents/Lightning-Booklet.pdf and https://lightningdev.umd.edu/aert/Safety.html)

Holle RL, Brooks WA, Cummins KL (2021) Lightning occurrence and casualties in U.S. National Parks. Wea Clim Soc 12:525–540

Kahl JDW (1998) National audubon society first field guide: Weather. Scholastic Inc., New York, p 159

Kramer S, Faidley W (1992) Lightning. Lerner Publications, Minneapolis, p 48

Krider EP, Ladd CG (1975) Upward streamers in lightning discharges to mountainous terrain. Weather 30:77–81

Ludlum DM (2001) Weather. Harper Collins Publishers, London, p 663

Ludlum DM, Holle RL, Keen RA (1995) The audubon society pocket guide: Clouds and storms. Alfred A Knopf, pp 192

Montanyà J, van der Velde OA, March V et al (2012) High-speed video of lightning and x-ray pulses during the 2009–2010 observation campaigns in northeastern Spain. Atmos Res 117:91–98

Parzybok TW (2005) Weather extremes of the west. Mountain press publishing company, Missoula, Montana, p 282

Saba MMF, Campos LZS, Krider EP et al (2009) High-speed video observations of positive ground flashes produced by intracloud lightning. Geophys Res Lett 36:L12811

Saba MMF, Schultz W, Warner TA et al (2010) High-speed video observations of positive lightning flashes to ground. J Geophys Res 115:D24201. https://doi.org/10.1029/2010JD014330

Salanave LE (1980) Lightning and its spectrum: An atlas of photographs. University of Arizona Press, Tucson, p 136

Schumann C, Hunt HGP, Smitt J et al (2022) Optical observations of 99 upward flashes in Johannesburg, South Africa. In: Preprints of the 36th international conference on lightning protection, Cape Town, South Africa, 02–07 October 2022.

Uman MA (1986) All about lightning. Dover Press, pp 167

van der Velde OA, Montanyà J, López JA et al (2019) Gigantic jet dischares evolve stepwise through the middle atmosphere. Nature Communications, pp 10

Wetmore RH II (2010) Thirty years of lightning photography in Southern Arizona. ISBN10:0982566212, pp 112

Wiewandt R, Wilks M (2001) The Southwest inside out: An illustrated guide to the land and its history. Wild Horizons Publishing Inc., Tucson, p 207

4

When, Where, and How Much Lightning Occurs in Arizona

Abstract Arizona is anything but uniform in terrain, so the topography strongly affects when, where, and how much lightning occurs. During the year, most lightning occurs from early July through September. It is also present mostly from late morning to early evening. Lightning tends to be concentrated in specific regions of the state, while there are wide variations in its pattern. This chapter will describe these differences in detail at the state, county, and city levels. The locations with the greatest lightning densities are in the eastern mountains and the least lightning densities are along the Colorado River in western Arizona. Year-to-year changes, and variations on the monthly and hourly time scales are also presented. Special attention is paid to lightning in the Grand Canyon, southeast mountains, Mogollon Rim, and western deserts.

4.1 The Monsoon in Arizona

The monsoon in Arizona is sometimes called the Arizona Monsoon or Southwest Monsoon within the U.S. However, it is part of the larger North American Monsoon, a seasonal change in wind direction on a large scale (Adams and Comrie 1997; Douglas et al. 1993). From the most basic perspective, the monsoon in Arizona is a shift from dry westerly winds to moist flow that is from the southeast to southwest for a few months. The North American Monsoon is not the only such system in the world. In fact, the Indian Subcontinent has by far the strongest monsoon in the world, when rain, thunderstorms, and lightning move across a large expanse of land

© Springer Nature Switzerland AG 2023
R. L. Holle and D. Zhang, *Flashes of Brilliance*,
https://doi.org/10.1007/978-3-031-19879-3_4

during the wet season, generally from south to north. For example, lightning occurrence and lightning fatalities in Bangladesh are concentrated from late April through early June during the pre-monsoon and early monsoon months (Holle et al. 2019). Similarly in North America, lightning activity starts to the south in Mexico during May and gradually makes its way north to reach Arizona by early July (Holle and Murphy 2015).

There are several cogent reasons for addressing the subject in Arizona separately from severe thunderstorms in other parts of the U.S. Among the main factors making this a special subject are the following:

- Arizona is on the northern edge of this much larger monsoon phenomenon that is strongest in northwest Mexico.
- This larger monsoon pattern is made possible by the warm sea surface temperatures in the Gulf of California (Sea of Cortez) and the tropical Eastern North Pacific Ocean that supply abundant low-level moisture (Jana et al. 2018).
- Tropical waves (easterly waves) arriving from the east or southeast travel from the Gulf of Mexico across Texas, Chihuahua, and New Mexico to provide moisture during the monsoon season, especially at middle levels of the atmosphere.
- Gulf surges originating in overnight mesoscale convective systems (MCS) in the southern Gulf of California can move northward and sometimes reach southwest Arizona; some gulf surges are triggered by tropical cyclones in the Pacific Ocean and Gulf of California (Higgins and Shi 2005).
- Moisture from tropical systems, named or unnamed, provides water vapor at lower levels of the atmosphere from the Eastern North Pacific and Gulf of California in remnants that are usually diffuse (Reyes and Mejia Trejo 1991; Corbosiero et al. 2009; Mejia et al. 2015).
- Occasional mid-latitude troughs and ridges brush Arizona from the north in weakened forms.
- One day's thunderstorm activity usually impacts the next day's storm development due to the cooling of the land from rain on the previous day.
- Upper air data can be sparse from stations in Mexico.
- Lightning activity occurs in bursts and breaks that often last for several days each throughout the monsoon season (Watson et al. 1994a).

A major improvement in understanding the impact of the North American Monsoon on Arizona has occurred because of several projects led in part by University of Arizona (UArizona) researchers. Nevertheless, daily forecasting of rainfall and lightning development during the monsoon months remains a challenge for Arizona due to the preceding list of possible factors. Several field projects have taken place with special surface and upper air dataset collection, starting with Southwest Area Monsoon Project (SWAMP) in 1993, then North American Monsoon Experiment (NAME) in 2004 (Farfán and Zehnder 1994; Higgins and Gochis 2007), and a series of other focused programs (Adams and Comrie 1997).

Personal Experience

The first author of this book spent a week as a day-shift forecaster for NAME 2004 and was intimidated by the variety of possible but subtle factors and how they interacted in unexpected ways to produce large amounts of thunderstorm activity.

Other current activities to understand the unique features of the monsoon in Arizona are being addressed by atmospheric scientists at UArizona. They are running a tailored numerical weather prediction model at very high resolution called the WRF (Weather Research and Forecasting). This model is being run with a resolution of 1.8 and 5.4 km such that it is the best model for studying the North American Monsoon. Results were posted in a discussion (http://arizonawrf.blogspot.com/) on many days during the monsoon. Additional activities at UArizona include monsoon outlooks at https://climas.arizona.edu/sw-climate/monsoon and podcasts at https://climas.arizona.edu/media/podcasts for periods longer than a day.

North American Monsoon lightning was explored with data from Vaisala's Global Lightning Dataset GLD360 network that uses similar technology to the National Lightning Detection Network (NLDN, Holle and Murphy 2015). An overview of the annual cycle of North American Monsoon lightning from this study in Fig. 4.1 shows its start in northwest Mexico in May that reaches Arizona in July, then slowly dissipates during September. Clearly, the monsoon lightning activity to the south over Mexico is greater than over Arizona, since the state is an extension of the peak activity to the south.

On a daily basis, Fig. 4.2 from Holle and Murphy (2015) shows maximum lightning from 2200 to 0200 UTC, which is 1500–1900 Mountain Standard Time (MST). That time period translates to 3:00 to 7:00 p.m. for most of Arizona. That is the year-around time zone for all of Arizona except the Navajo Reservation that observes Mountain Daylight Time in the summer months to match neighboring states to the north and east of Arizona. More details are given for both monthly and hourly time changes in the following sections of this chapter.

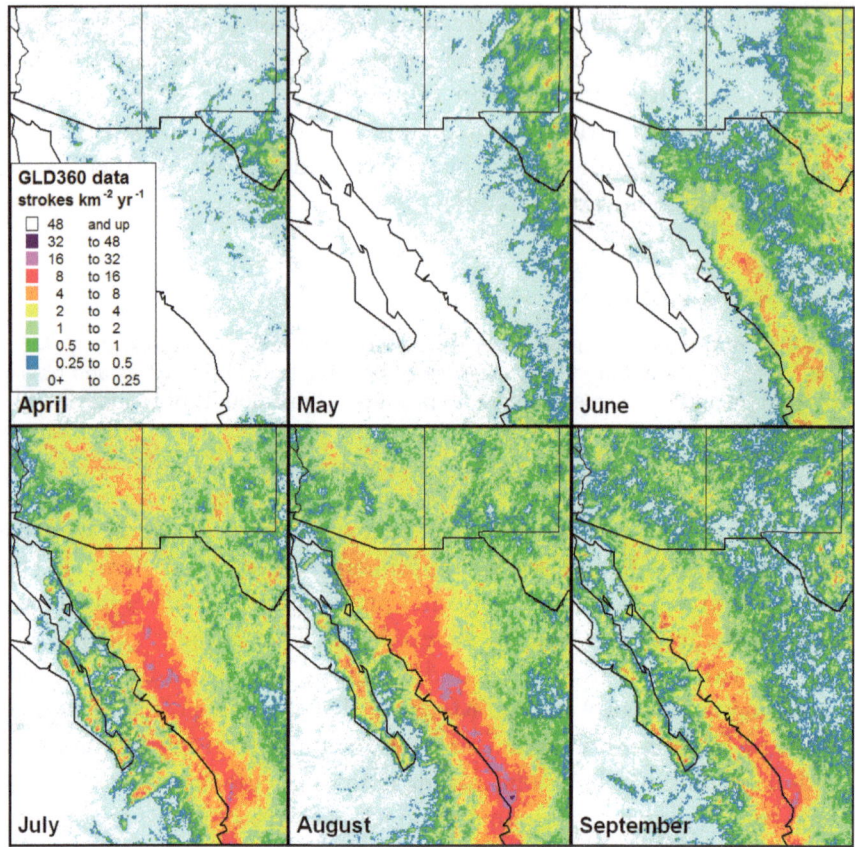

Fig. 4.1 Monthly maps from April to September, 2011 through 2014, of lightning frequency over the monsoon area of southern Arizona and northwestern Mexico (Holle and Murphy 2015). © American Meteorological Society. Used with permission

Time Conversion

Meteorologists use UTC in their internal operations and publishing research since it does not depend on the local time zone. For example, the maps in Figs. 4.1 and 4.2 include two time zones, and some areas use only standard time and some use daylight time in the summer monsoon months. UTC stands for Coordinated Universal Time and is the same as GMT (Greenwich Mean Time).

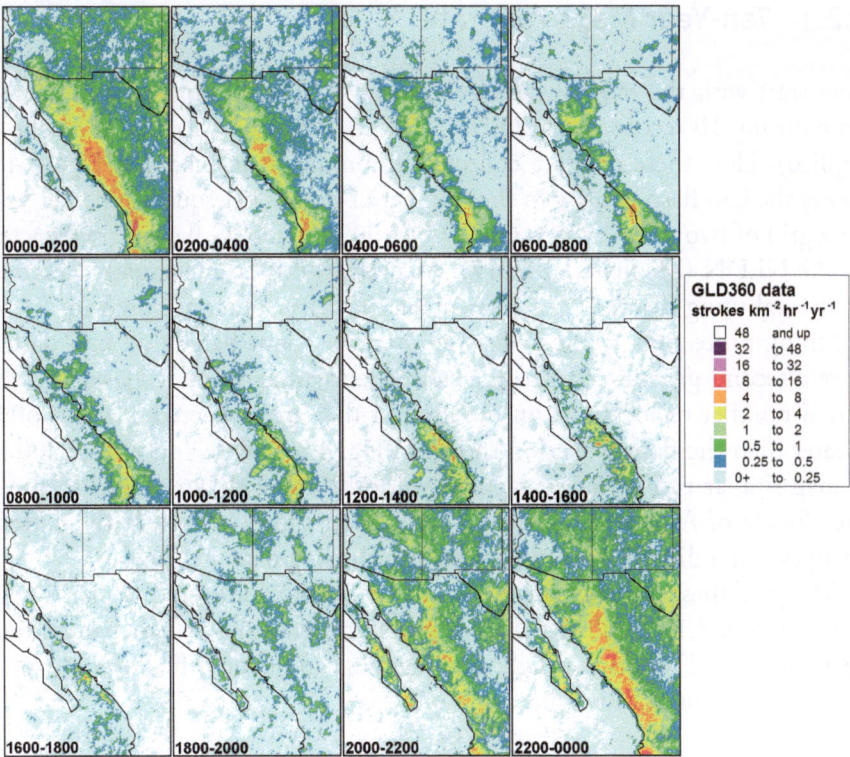

Fig. 4.2 Two-hour maps of lightning frequency over the monsoon area of southern Arizona and northwestern Mexico from April to September, 2011 through 2014 (Holle and Murphy 2015). Time is in UTC; subtract 7 h for Mountain Standard Time. © American Meteorological Society. Used with permission

4.2 Lightning Occurrence by Area in Arizona

Where is the most lightning in Arizona? Where is the least? Why are these high and low areas where they are? What time of year? What time of day? How much lightning contacts the ground in a typical square mile?

These questions about Arizona lightning have been asked repeatedly by the media in print, broadcast, and digital forms throughout the state. As a result, an inventory of maps and tables had been developed that is assembled in this chapter for a matching time period. Similar information has been posted in various formats for more than 20 years on websites of the National Weather Service at https://www.weather.gov/safety/lightning, National Lightning Safety Council at http://lightningsafetycouncil.org/LSC-Resources.html, and Vaisala at Vaisala Annual Lightning Report.

4.2.1 Ten-Year Maps for State

Let's start with the map of cloud-to-ground (CG) lightning over the entire state during 10 recent years (2009–2018, no 15-kA positive cutoff has been applied). How these data are collected is considered in Chap. 6. Figure 4.3 shows the CG flash density in the number of flashes per square mile per year on a grid of two miles by two miles. Over 90% of all CG flashes are detected by the NLDN (Medici et al. 2017). As a result, this is an accurate depiction of the lightning distribution in terms of CG flashes that are used for many lightning protection applications. On average, about half of all CG flashes have a second ground contact in a subsequent stroke (Sect. 1.4), so that the actual number of locations on the ground that are impacted is about 50% greater than the flash depiction shown in Fig. 4.3 (Valine and Krider 2002). A map similar to Fig. 4.3 was published for 2004–2013 in an article about the climate of Arizona (Holle et al. 2015) that shows how little the pattern changes when differing multiple years of lightning data are used.

Many of these lightning features can be readily connected to the topography in Fig. 3.1. Recall that Fig. 3.1 showed lightning over the contiguous 48 U.S. states. Five distinct lightning regimes can be identified over Arizona, starting from south to north:

- Southeast: This region, including Tucson, has frequent lightning over the mountains in the basin-and-range geography. This area is the northern extension of a much larger lightning maximum over the Sierra Madre Occidental mountains of northern Mexico (Holle and Murphy 2015).
- Southwest: This low-elevation region, including the western side of the Phoenix area, has a very low occurrence of lightning. It is least on the California border to the west.
- East-central border: This area of the White Mountains is within a large region at high altitudes and has a broad lightning maximum that extends into New Mexico to the east in the U.S. map in Fig. 3.1. This region is where many of the Apache lightning anecdotes in Sect. 2.2.3 are based.
- Mogollon Rim: A moderate lightning frequency is in this area due to a sharp terrain gradient extending from the White Mountains in the east toward the Nevada border on the northwest; this is called the Rim country.
- Northeast Plateau: A lightning minimum is over this high-elevation region that has few mountains to provide terrain gradients for initiating thunderstorms. On some monsoon-season days, lightning forms over the Chuska

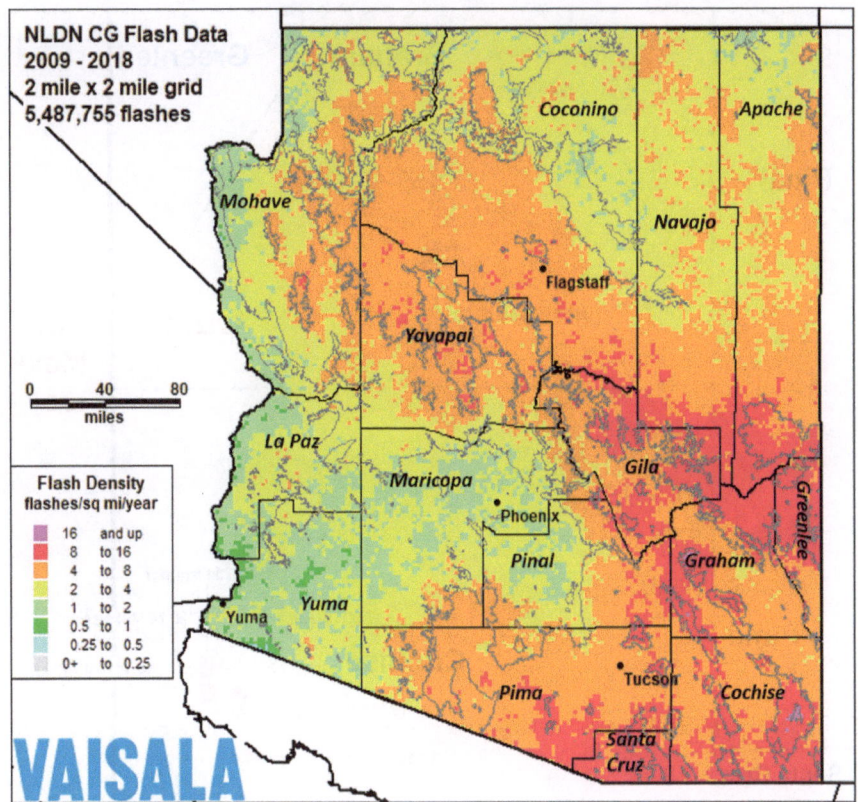

Fig. 4.3 Density of CG flashes per square mile per year on a grid of 3.219 by 3.219 km (two miles by two miles) over Arizona based on a total of 5,487,755 CG flashes from the NLDN from 2009 through 2018. County names are in italics; four cities are indicated with dots. County lines are defined by black lines, and state borders by thick lines. Scale in middle left (*Courtesy* Vaisala Inc.)

Mountains to the north of the Rim along the New Mexico border (northeast edge of Fig. 2.1); these storms sometimes reach the Rim to the south and enhance activity there.

Where are the grid squares with the most CG lightning in this 10-year map? Figure 4.4 shows that all locations of the top ten grids are in the southeast where small pink squares are visible in Fig. 4.3. There are three groups in Fig. 4.4:

- The largest is 20.8 CG flashes per square mile per year on Chiricahua Peak in the Chiricahua Mountains. Five other grid squares in the top ten are also in this mountain range that includes Chiricahua National Monument in Cochise County.

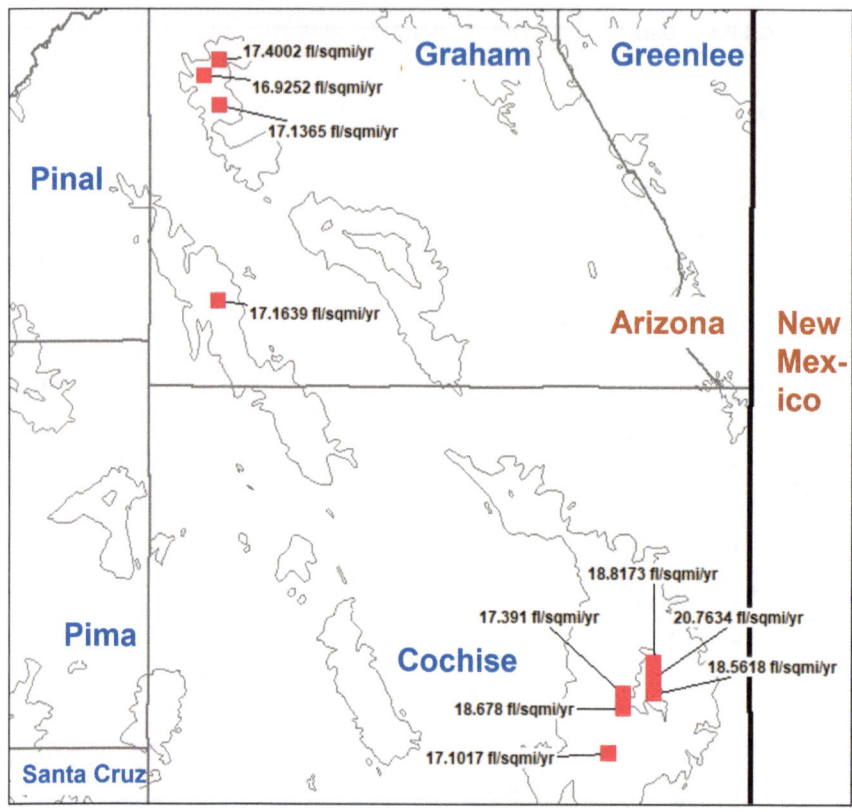

Fig. 4.4 Map of the top ten densities of CG flashes (shown by pink squares) per square mile per year over Arizona on a grid of 3.219 by 3.219 km (two miles by two miles). County names in blue and state names in orange. County lines indicated by thin black lines, and New Mexico border by thick black line. NLDN data are from 2009 through 2018 (*Courtesy* Vaisala Inc.)

- To the north, three more top ten grid squares with large CG flash densities are in the San Carlos Apache Indian Reservation near Mt. Turnbull in the Santa Teresa Mountains of Graham County.
- One top ten grid square is a few miles north of Bassett Peak in the Galiuro Mountains of Graham County.

Persistent Storm Locations

It's not so much that the Chiricahua and White Mountain regions have more lightning in each storm. Instead, day after day, they have lightning when other areas don't, so the total for the year accumulates to produce local maxima.

Where are the grid squares with the least CG lightning in Arizona? Figure 4.5 shows these locations to be along the northwest and southwest borders of the state. There are three groups:

- Two grid squares have zero CG flashes in the 10-year dataset. One is in the city of San Luis in the very southwest corner of Arizona in Yuma County. The other no-flash grid square is near Bonelli Landing of Lake Mead on the border with Nevada.

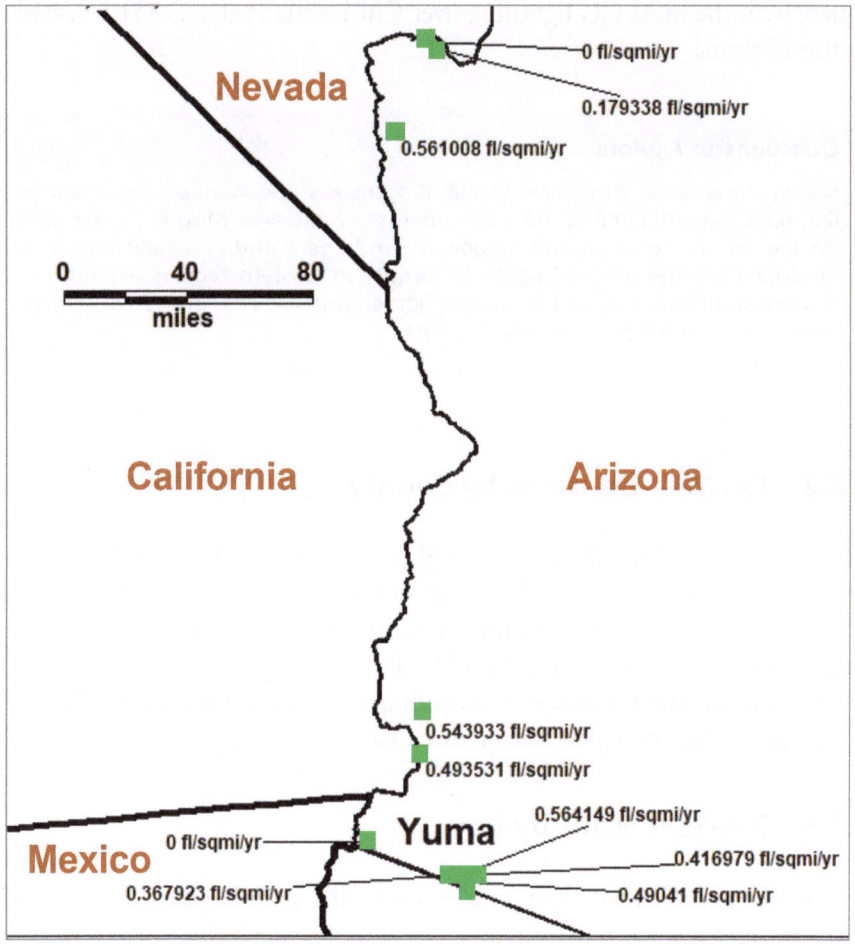

Fig. 4.5 Map of the lowest ten densities of CG flashes (shown by green squares) per square mile over Arizona per year on a grid of 3.219 by 3.219 km (two miles by two miles). Labels show state names and location of Mexico in orange, city of Yuma in black, and state and international borders by thick black lines. NLDN data are from 2009 through 2018 (*Courtesy* Vaisala Inc.)

- Six more of the bottom ten grid squares are in Yuma County. Four are in low uninhabited desert areas near the Mexican border, and two in remote areas along the Colorado River.
- The other two are in the northwest corner of the state in Mohave County. One is adjacent to the zero value at Lake Mead, and the other is at a low altitude in unoccupied rugged terrain east of the Colorado River.

For perspective, the grid square with the tenth smallest density has 22 CG flashes during 10 years; this is along the Mexican border. In contrast, the grid square with the most CG lightning over Chiricahua Peak has 851 CG flashes in the 10 years!

Questions to Explore

Seeing these large differences in lightning makes one wonder about nearby lightning. Does the Chiricahua maximum extend into New Mexico and Mexico? Do the White Mountain and Mogollon Rim large densities extend into New Mexico? Does the minimal lightning density in western regions extend into Nevada and California, and across the border with Mexico? The GLD360 maps in Figs. 4.1 and 4.2 provide some insights.

4.2.2   Ten-Year Densities by County

The same data in Fig. 4.3 are now graphed by county in Fig. 4.6. The highest county-wide CG density is 9.33 CG flashes per square mile per year in Santa Cruz County on the Mexican border south of Tucson. Other southern and eastern counties are also ranked highly. The lowest CG density by county is in the southern and western deserts, ending with the least density of 1.79 CG flashes per square mile per year in Yuma County.

4.2.3   Ten-Year Maps by City

Ninety percent of the population of Arizona, the sixth-largest U.S. state, lives in urban areas, since much of the state is not suitable for small-scale agriculture. In addition, 82% of the state is controlled by tribal, federal, or state entities in the form of National Parks, National Monuments, Indian Reservations, Department of Defense facilities, Fish and Wildlife Refuges, National Forests, or maintained by the Bureau of Land Management.

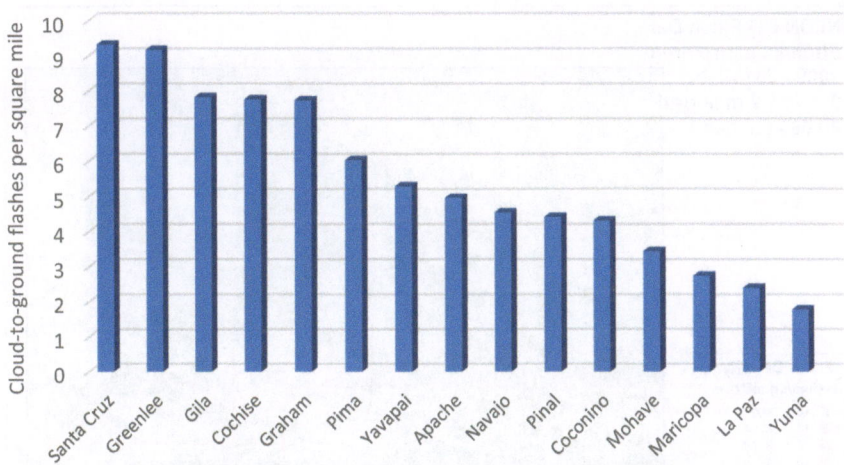

Fig. 4.6 County densities of CG flashes per 2.59 square kilometers (one square mile) per year from 2009 to 2018, ranked in order of density (*Courtesy* Vaisala Inc.)

Focusing on the most populated areas, the following figures use the same color scale as Fig. 4.3 to show a more detailed representation of CG lightning density for the four major cities and the metropolitan areas surrounding them. In order of population, these are

- Phoenix metropolitan area: This region (Fig. 4.7) extends outward to the mountains on the north and east sides and reaches into the lower deserts to the west and south (elevation map in Fig. 2.1). The CG lightning activity is quite weak throughout, and only a hint of a larger density is apparent over the mountains to the northeast. Otherwise, the spatial variations are due to individual storms that occurred within each grid square during the 10-year period.
- Tucson metropolitan area: This region (Fig. 4.8) extends outward to large nearby mountains on all but the southwest through northwest sides (elevation map in Fig. 2.1). The lightning activity is much stronger than in Phoenix, and the map includes large densities over the mountains in many directions.
- Flagstaff metropolitan area: This region (Fig. 4.9) has a larger flash density than Phoenix and is about the same as for Tucson. Flagstaff is situated within the broad lightning maximum of the Mogollon Rim region of Fig. 4.3 that extends across the area. The highest mountains of the San Francisco Peaks account for the maximum to the north of the city in Fig. 4.9. To the east is shown the large-scale decrease in lightning that includes the northeast plateau (elevation map in Fig. 2.1).

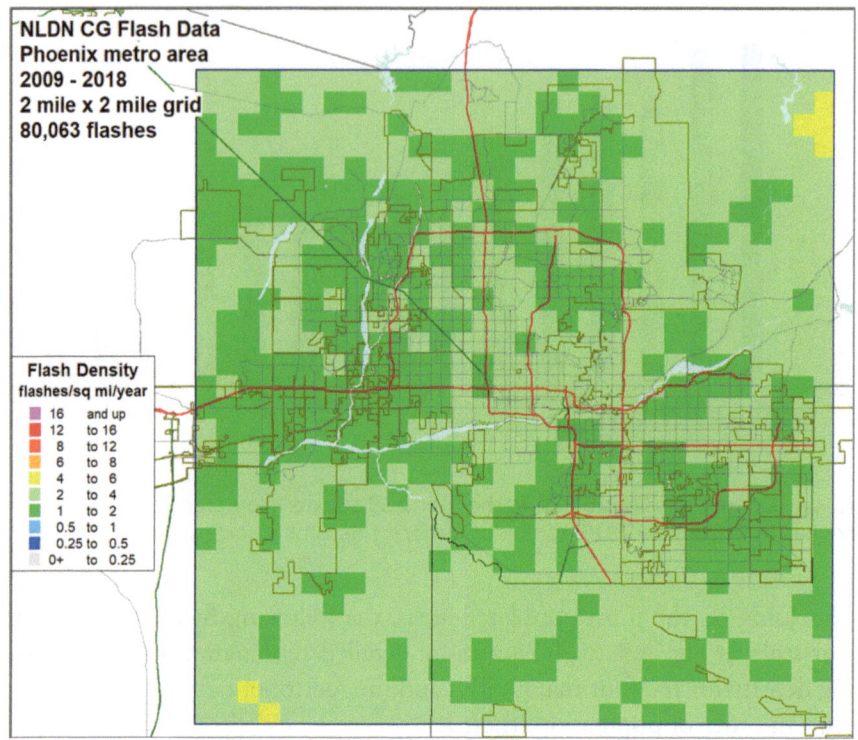

Fig. 4.7 CG flash densities over the Phoenix metropolitan area per square mile per year on a grid of 3.219 by 3.219 km (two miles by two miles) based on 80,063 flashes from the NLDN from 2009 through 2018 (*Courtesy* Vaisala Inc.)

- Yuma metropolitan area: This region (Fig. 4.10) has small CG flash densities throughout, including some zero and near-zero grid squares as also indicated in Fig. 4.5. Lightning here is a special occasion!

4.2.4 Comparisons with Other States and Countries

Let's take a step back from Arizona to be sure to understand the context of our lightning density. Figure 3.1 shows the CG flash density for the U.S. in the same color scale as for Arizona in Fig. 4.3. To place the state into context, Table 4.1 lists the CG flash densities of the 48 contiguous states and the District of Columbia for the same time period. In Table 4.1, the CG flash density for Arizona is listed as an average of 4.7 flashes per square mile (12.2 flashes per square kilometer) and ranks number 33 in the state listing. The flash density is almost the same as its Colorado neighbor to the northeast that is ranked number 32. Similar tables for lightning occurrence and fatalities are

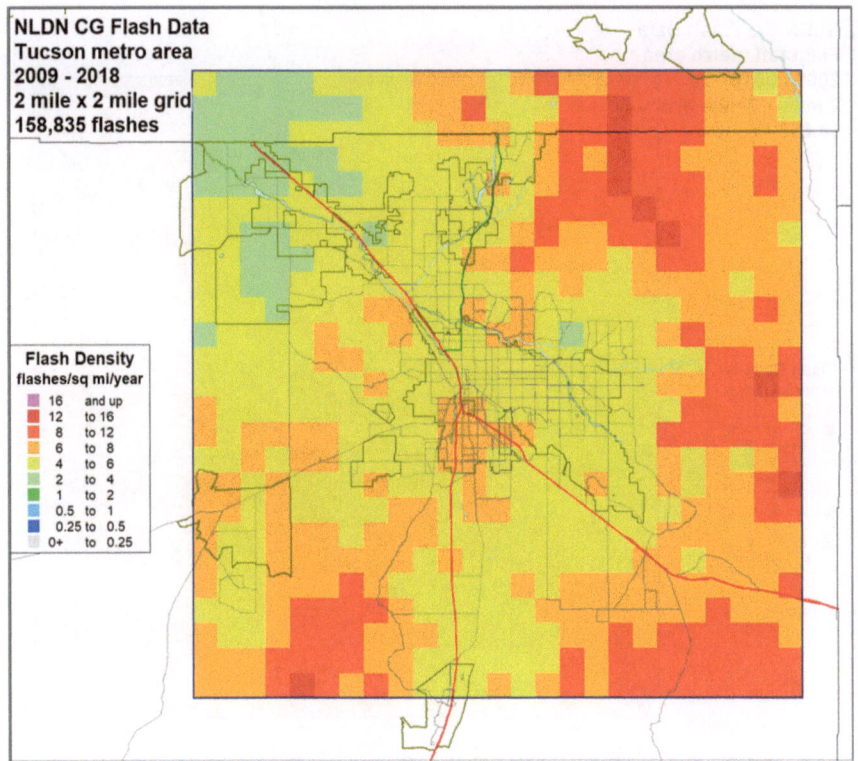

Fig. 4.8 CG flash densities over the Tucson metropolitan area per square mile per year on a grid of 3.219 by 3.219 km (two miles by two miles) based on 158,835 flashes from the NLDN from 2009 through 2018 (*Courtesy* Vaisala Inc.)

updated in various formats on the websites of the National Weather Service at https://www.weather.gov/safety/lightning-science and the National Lightning Safety Council at http://lightningsafetycouncil.org/.

An exact comparison of Arizona lightning with the rest of the world is not available. Nevertheless, one source to consider is data from the Global Lightning Dataset GLD360 network (Said et al. 2010; Said 2017). Figure 4.11 shows the global stroke density from 2014 to 2018, a period that includes the last half of the years shown in many other figures in this chapter; the GLD360 has been developed and deployed by Vaisala. This dataset is not calibrated to provide CG flashes specifically but indicates relative values around the world. One can see that Arizona is in the range of 8–16 strokes per square kilometer per year, shown in red on this scale. A stroke density between 8 and 16 is not unusual over landmasses over all continents but Antarctica. Another feature to note in Fig. 4.11 is the much greater stroke density a short distance

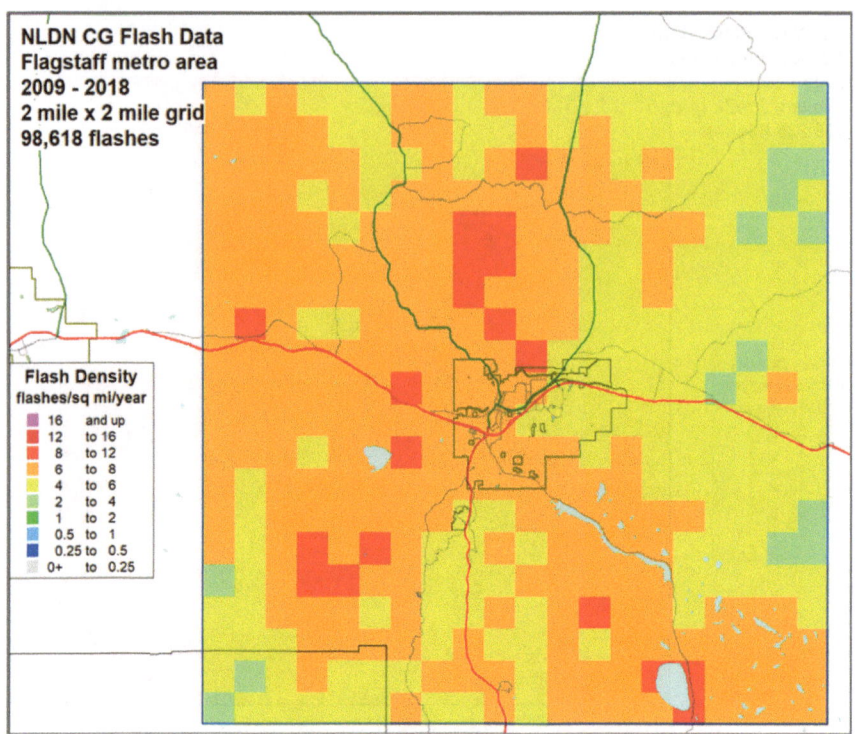

NLDN CG Flash Data
Flagstaff metro area
2009 - 2018
2 mile x 2 mile grid
98,618 flashes

Flash Density
flashes/sq mi/year

16	and up	
12	to	16
8	to	12
6	to	8
4	to	6
2	to	4
1	to	2
0.5	to	1
0.25	to	0.5
0+	to	0.25

Fig. 4.9 CG flash densities over the Flagstaff metropolitan area per square mile per year on a grid of 3.219 by 3.219 km (two miles by two miles) based on 98,618 flashes from the NLDN from 2009 through 2018 (*Courtesy* Vaisala Inc.)

south of Arizona in Mexico in association with the North American Monsoon (Adams and Comrie 1997; Douglas et al. 1993; Holle and Murphy 2015).

The GLD360 map shows approximately the same extremes in lightning density around the world as satellite data (Albrecht et al. 2016; Zipser et al. 2006). However, there is not yet a global CG flash density climatology using calibrated data. While the patterns are certainly similar, it is not as clear where the maxima will turn out to be located. The main source of differences is the amount of in-cloud lightning pulses that are a greater portion of the lightning frequency in some places than others. This is an active area of research at several facilities around the world such as by Medici et al. (2017).

4.2.5 Arizona Lightning Mapping from Satellites

Another view of lightning in Arizona can be made from satellite sensors rather than ground-based networks. Figure 4.12 shows the total lightning density

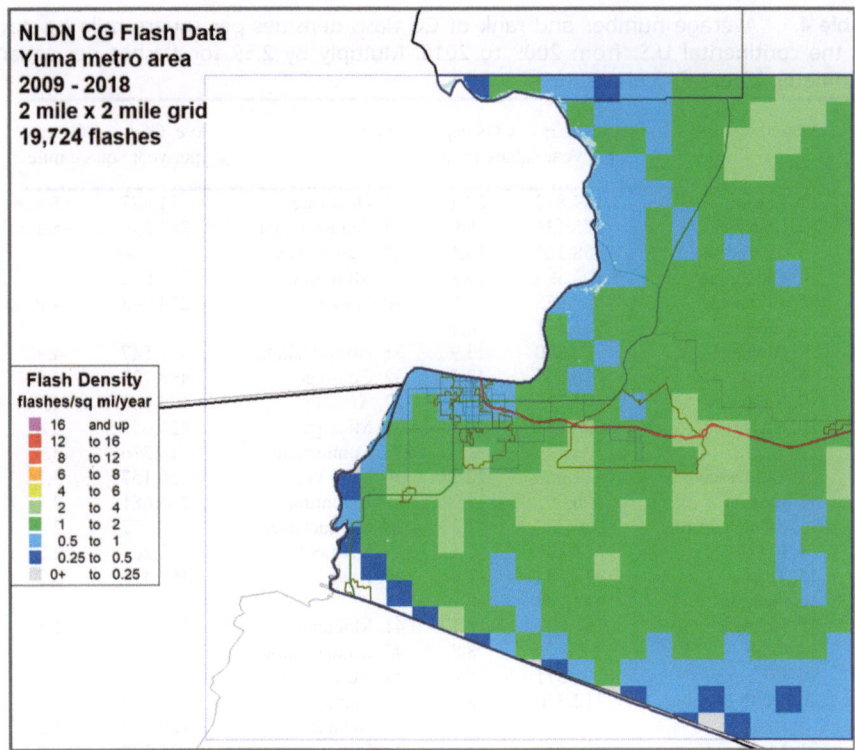

Fig. 4.10 CG flashes over the Yuma metropolitan area within Arizona per square mile per year on a grid of 3.219 by 3.219 km (two miles by two miles) based on 19,724 flashes from the NLDN from 2009 through 2018 (*Courtesy* Vaisala Inc.)

map for Arizona on a 0.05° by 0.05° grid detected by the Lightning Imaging Sensor (LIS) on the Tropical Rainfall Measuring Mission (TRMM) satellite during 2004–2013. Unlike the NLDN, LIS detects a narrow near-infrared band of lightning emissions from the cloud top. The spatial resolution of TRMM-LIS was about 4 km by 4 km at nadir and somewhat larger at the edges of the field of view. Due to the orbiting nature of the satellite (16 orbits per day), TRMM-LIS only observed at a certain place on earth for about 90 s (Zhang et al. 2019). The overall spatial distribution looks similar to what NLDN shows in Fig. 4.3. However, there are some differing hotspots in the map, primarily due to LIS's limited sampling duration. Note that LIS detects total lightning including both CG strokes and in-cloud pulses, which is different from CG flashes only in Fig. 4.3. A more detailed description of satellite-based lightning observation and related studies will be discussed in Chap. 7.

Table 4.1 Average number and rank of CG flash densities per square mile by state in the continental U.S. from 2009 to 2018. Multiply by 2.59 for flashes per square kilometer (*Courtesy* Vaisala Inc.)

State	Ave. CGs per year	CGs per square mile	State	Ave. CGs per year	CGs per square mile
1. Florida	1,189,873	20.8	26. Delaware	11,487	5.8
2. Louisiana	875,136	18.9	27. Pennsylvania	241,330	5.3
3. Mississippi	768,126	16.1	28. New Jersey	40,286	5.3
4. Oklahoma	1,102,399	15.8	29. Minnesota	425,932	5.0
5. Arkansas	837,978	15.7	30. Wisconsin	274,963	4.9
6. Missouri	981,351	14.1			
7. Alabama	720,780	13.9	31. North Dakota	342,547	4.9
8. Illinois	716,885	12.7	32. Colorado	496,722	4.8
9. Kentucky	502,715	12.5	33. Arizona	539,689	4.7
10. Kansas	1,023,177	12.5	34. Michigan	220,635	3.8
			35. Connecticut	16,316	3.3
11. Tennessee	511,650	12.2	36. New York	150,157	3.1
12. Indiana	406,902	11.3	37. Wyoming	265,681	2.7
13. South Carolina	346,282	11.2	38. Massachusetts	19,177	2.4
14. Iowa	625,437	11.1	39. Vermont	21,261	2.2
15. Texas	2,923,584	11.0	40. Utah	187,410	2.2
16. Georgia	642,203	10.9			
17. Nebraska	780,934	10.1	41. Montana	308,166	2.1
18. Ohio	365,476	8.8	42. Rhode Island	2,210	2.1
19. D.C.	571	8.4	43. New Hampshire	17,238	1.9
20. North Carolina	412,340	8.3	44. Maine	40,380	1.2
			45. Nevada	129,580	1.2
21. Maryland	73,218	7.4	46. Idaho	79,327	1.0
22. S. Dakota	540,604	7.0	47. California	81,689	0.5
23. Virginia	271,175	6.8	48. Oregon	46,190	0.5
24. West Virginia	159,353	6.6	49. Washington	24,490	0.4
25. New Mexico	731,776	6.0			

4.3   Lightning Occurrence by Time in Arizona

4.3.1   Year-to-Year Variability

The average annual number of CG flashes in Arizona during the 10-year period is 561,835 (Fig. 4.13). These NLDN data show a variation from 399,885 flashes statewide in 2009 to a maximum of 700,705 in 2015. In this 10-year sample, the largest year has 75% more lightning than the year with the least. Other years may have larger or smaller totals than shown in this 10-year sample. While the NLDN had some upgrades at various times through these years (Murphy et al. 2021), most of these changes in annual lightning counts can be attributed to varying meteorological situations.

One of the main reasons for the variability from year to year is the presence or absence during an individual year of large lightning events. Figures 1.15

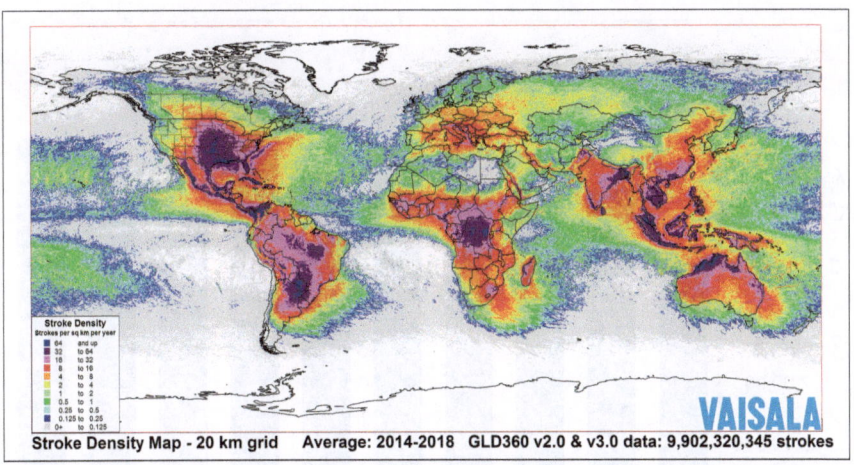

Fig. 4.11 Lightning stroke density over the globe in a grid of 20 km by 20 km based on a total of 9,902,320,345 strokes from Vaisala's Global Lightning Dataset GLD360 network from 2014 through 2018. © American Meteorological Society. Used with permission

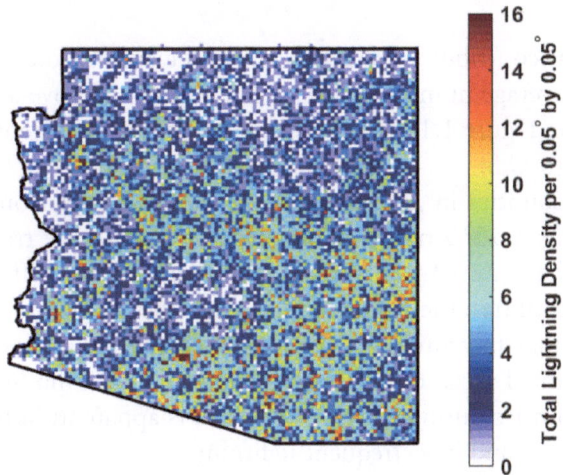

Fig. 4.12 Total lightning density on a grid of 0.05° by 0.05° in Arizona during 2004–2013, detected by the TRMM-LIS satellite

and 1.16 showed satellite views of mesoscale convective systems (MCS) that were among the largest such storm complexes in North America at the time. In terms of lightning, Fig. 4.14 shows a 10-h overnight event in August 2019 that had 142,541 combined CG strokes plus in-cloud events over the southern portion of the state; some additional lightning events detected to the north within the state are not shown. About 20% of these events were

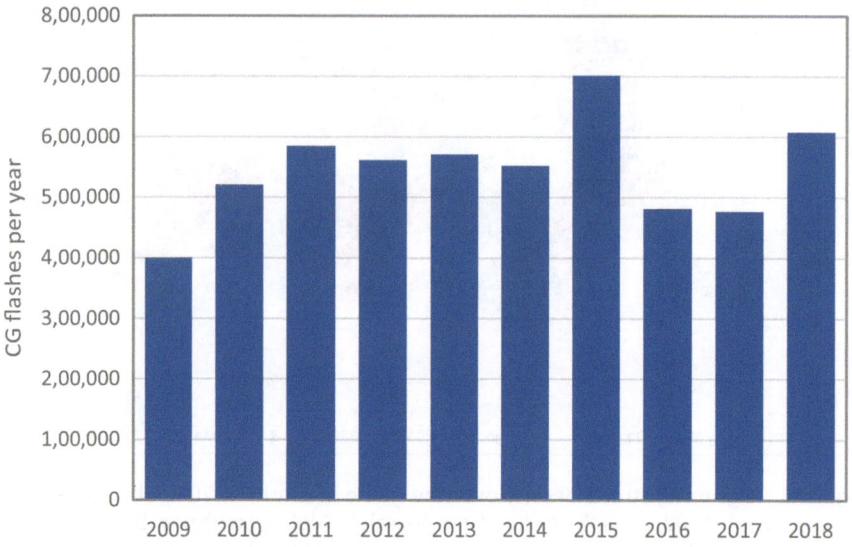

Fig. 4.13 CG flash counts by year over Arizona detected by the NDLN from 2009 to 2018 (*Courtesy* Vaisala Inc.)

CG flashes that contribute several percent to the annual total, so a year with more than an average number of MCSs can lead to an above-average year for the whole state. Figure 1.16 shows a satellite view of the extent of this MCS over the western U.S.

Do all the counties in Arizona have the same changes from year to year as shown in Fig. 4.13? An indication that this is generally true is shown in Fig. 4.15 for the same 10-year period. Although there are differences among the years and counties, the whole state tends to increase or decrease together, an indication that meteorological processes larger than Arizona are responsible for the annual variations. The dominance of Coconino (gray) and Pima (light blue) counties during many of the years is apparent; both counties are quite large in area and have frequent lightning.

4.3.2   Month-to-Month Variability

The strong concentration of Arizona lightning during the southwest monsoon accounts for most of the year's activity. That remarkable period will be shown in this section and discussed by region in Sect. 4.4. Monthly changes through the year in Figs. 4.1 and 4.16 show an extreme concentration during the few months of the monsoon. The same data in Table 4.2 show the number of flashes by month. Note two major results in this table:

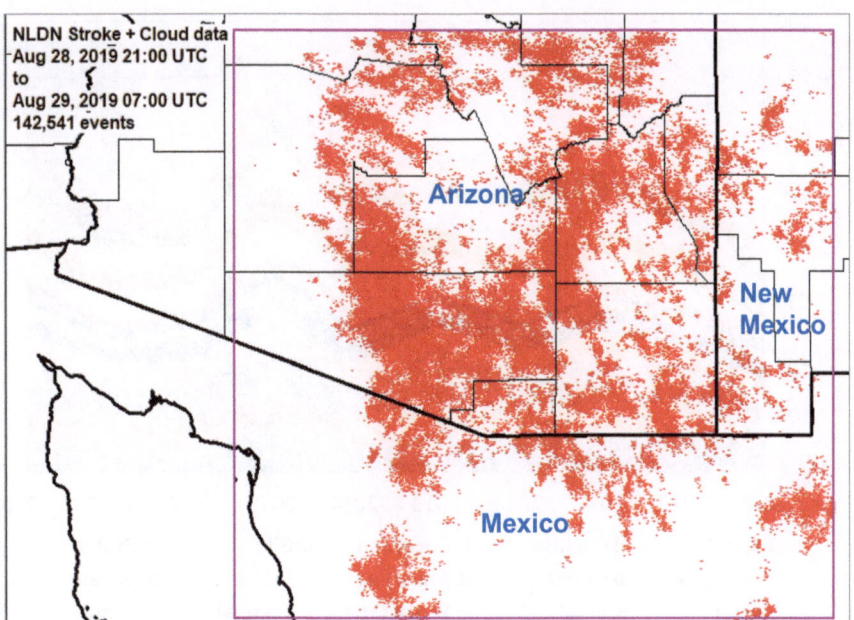

Fig. 4.14 Map of 142,541 CG strokes plus cloud pulses over the southern half of Arizona during a 10-h period from 2100 UTC (2:00 p.m. MST) on August 28 to 0700 UTC (midnight MST) on August 29, 2019. The magenta box represents the area of analysis (*Courtesy* Vaisala Inc.)

- The monsoon-season months of July, August, and September account for 89.8% of the year's lightning.
- Other months have few or almost no flashes. (Note: In fact, 2 of the 10 years have no flashes at all in November detected by the NLDN over Arizona.)

The concentration of lightning from July through September is very robust. Figure 4.17 shows few yearly variations from the annual cycle in Fig. 4.16 and Table 4.2. Notice that larger spikes in lightning occur in June and October of some years. Most worrisome is the occasional flare-up of a dry thunderstorm in June, the hottest and driest month in the state. In early June 2020, a dry thunderstorm produced a few CG flashes in southern Arizona and started the Bighorn Fire that burned throughout June because no rain-producing storms followed it for several weeks (Sect. 5.3.9).

City-by-city changes by month are shown in Fig. 4.18; these counts are for rectangles including the major population regions of the cities within Arizona shown in Figs. 4.7, 4.8, 4.9, and 4.10. The following are major results from this monthly tabulation:

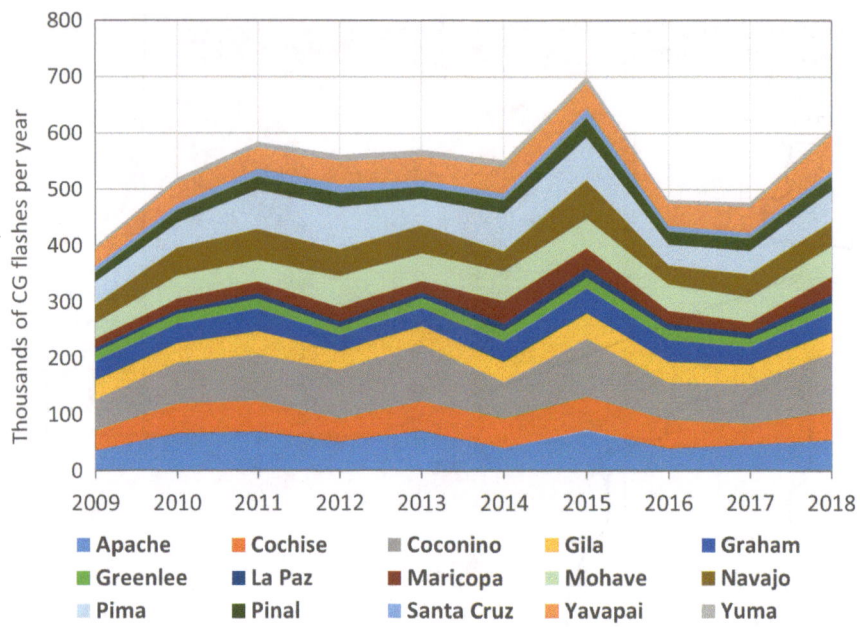

Fig. 4.15 CG flash counts per county by year over Arizona from 2009 to 2018 (*Courtesy* Vaisala Inc.)

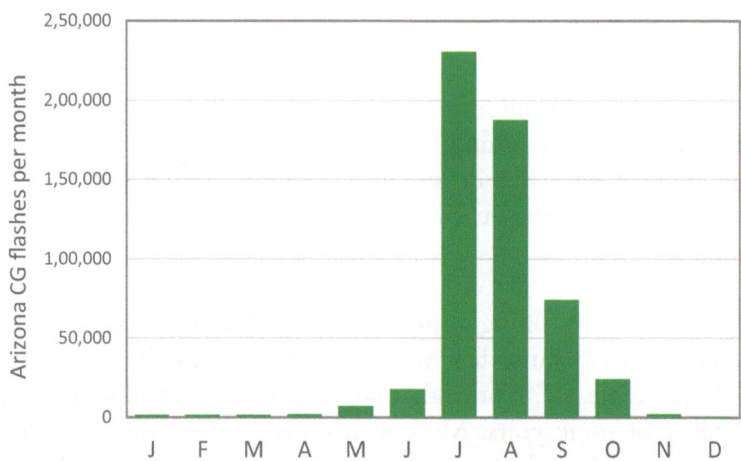

Fig. 4.16 CG flash counts by month per year over Arizona from 2009 to 2018 (*Courtesy* Vaisala Inc.)

Table 4.2 Average annual number of CG flashes by month, and percent of the annual total across Arizona from 2009 to 2018 (*Courtesy* Vaisala Inc.)

Month	Average flashes per year	Percent of annual total
January	584	0.1
February	646	0.1
March	1,506	0.3
April	1,860	0.3
May	7,221	1.3
June	17,818	3.2
July	230,804	42.1
August	187,628	34.2
September	74,061	13.5
October	24,055	4.4
November	2,127	0.4
December	466	0.1
Annual	548,776	

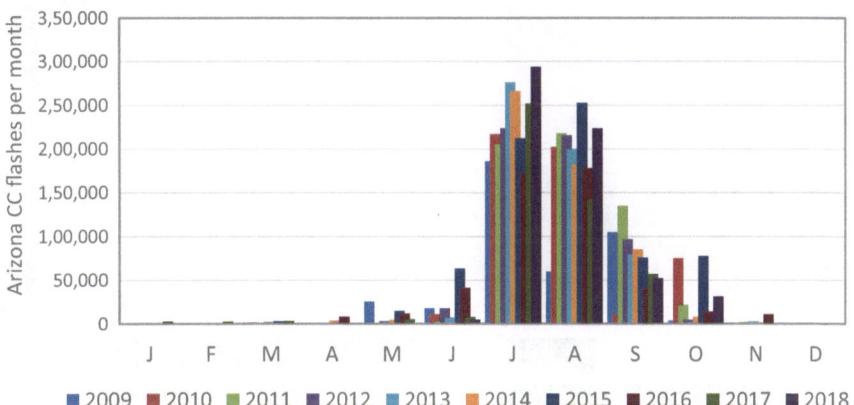

Fig. 4.17 CG flashes by month over Arizona for each year from 2009 to 2018 (*Courtesy* Vaisala Inc.)

- Flagstaff and Tucson have the same features as for the entire state in Fig. 4.16 and Table 4.2. These are a rapid increase from June to July and August, then a steady decrease into September and October.
- Phoenix lightning starts later in July and reaches a maximum in August, since monsoon activity often does not reach there from the south until later in July.
- Yuma has a later lightning season since the monsoon has difficulty reaching that far west from its origins in southeast Arizona, Mexico, and New Mexico. Instead, a different factor is at play for Yuma that results in up to half of its summer rainfall occurring in August and September. Many

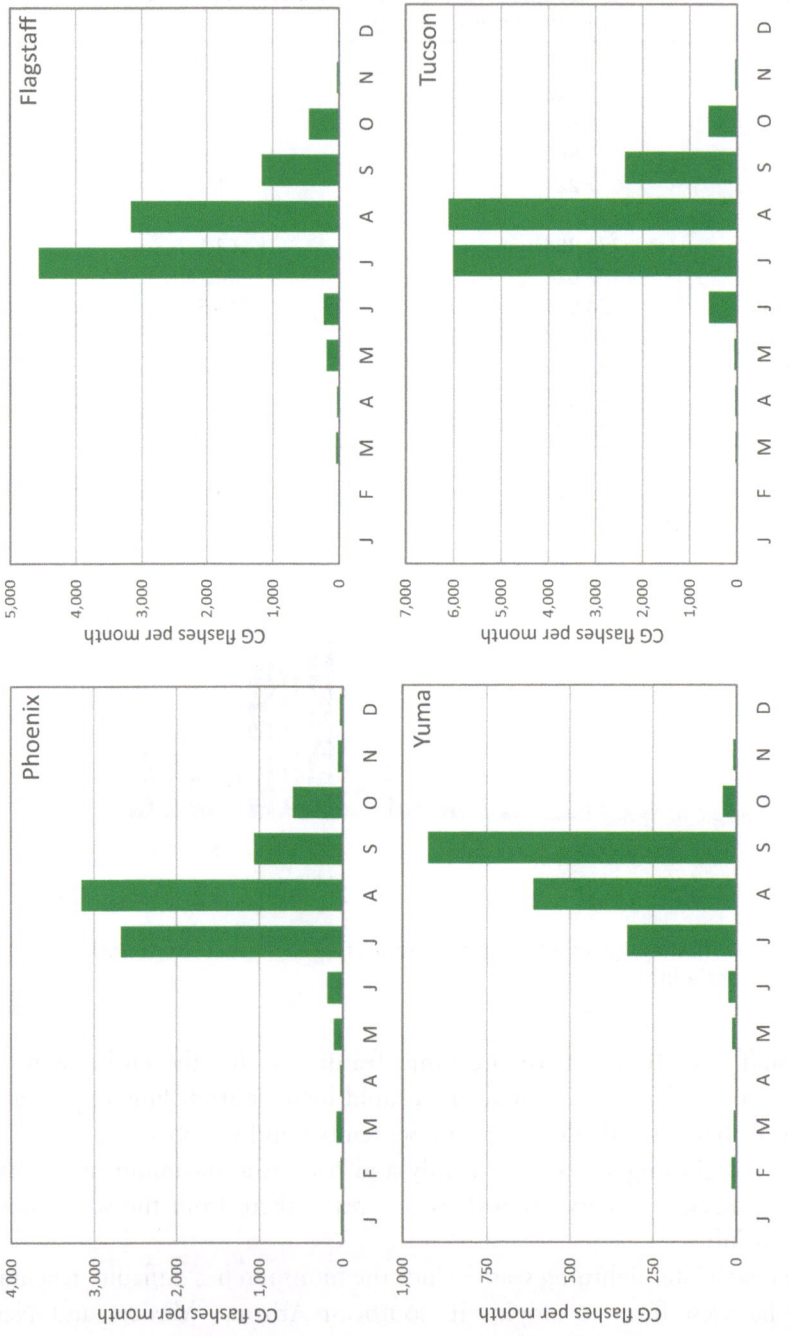

Fig. 4.18 CG flash counts by month per year for four major cities in Arizona from 2009 to 2018 (*Courtesy Vaisala Inc.*)

of these CGs come from active or decaying tropical depressions and tropical storms that arrive from the south or south-southwest, often associated with Gulf of California surges (Reyes and Mejia Trejo 1991; Higgins and Shi 2005; Corbosiero et al. 2009; Adams et al. 2014).

During the monsoon, lightning tends to make ground contact over the higher mountain peaks, the Mogollon Rim, southeast Arizona mountain ranges, and other large elevation areas as shown in Fig. 4.3 and described in Sect. 4.2.1. However, lightning in winter thunderstorms often is not affected by topography. A nighttime thunderstorm in southeast Arizona in December 2019 (Fig. 4.19) was caused by an upper level low. The low had been moving down the U.S. west coast until its trough axis between Tucson and San Diego was moving very slowly from west to east. Thunderstorms broke out after midnight and produced both in-cloud and CG lightning. Notice that no line or cluster is apparent in the pattern, only a scattering of events, indicating that neither topography nor a thunderstorm steadily moving along a track was involved. Lightning in this distributed pattern was formed by processes aloft due to a larger scale feature, in this case an upper low on the scale of several states. Local topography did not play an important role in where the lightning occurred.

Nighttime Storm

An additional clue that a larger scale factor was important here is that lightning was during the nighttime, so it was not affected by local daytime heating of the mountains in this area.

4.3.3 Time-of-Day Variability

Since most of the year's lightning in Arizona occurs during the monsoon season, and is most frequent over and near large mountain ranges and escarpments, is all the lightning only during the afternoon due to daytime heating? This section will explore the hourly (diurnal) change in lightning for the state and major cities; Watson et al. (1994b) also made a previous review.

For all of Arizona, Fig. 4.20 shows a strong maximum in the afternoon. The smallest hourly amount of CG flashes is from 0800 to 1000 MST. Then the number of flashes doubles every hour until slowing to the peak in the hour starting at 1500 MST (3:00 p.m.). Activity then slowly trends downward through the evening hours, but not as quickly as the rise in the late

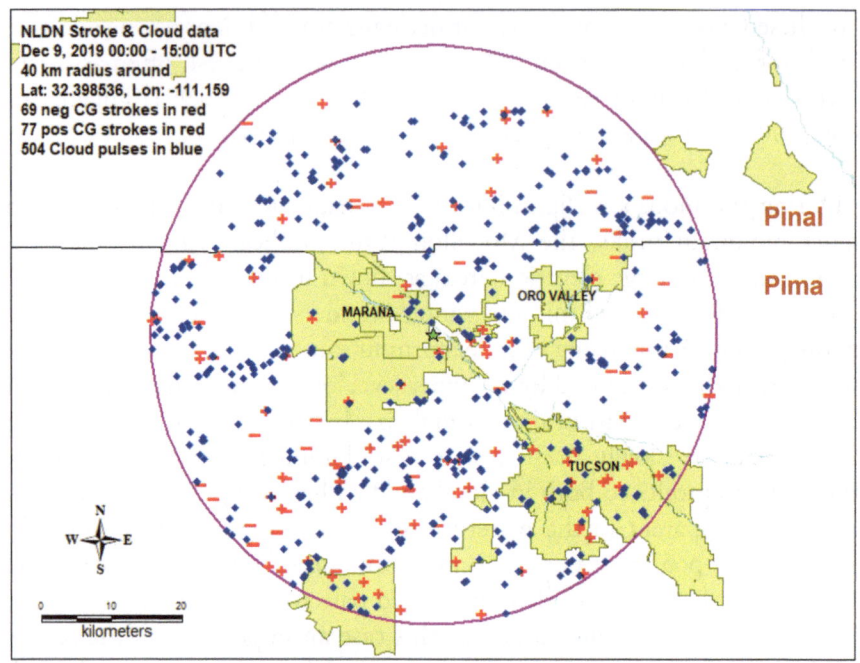

Fig. 4.19 CG strokes (red) and cloud pulses (blue) over southeast Arizona for 15 h on December 9, 2019. Time period is from 00:00 to 15:00 UTC, which is 5:00 p.m. MST on December 8 through 8:00 a.m. MST on December 9. CG strokes lowering negative charge are shown by minus signs, CG strokes lowering positive charge by plus signs, and cloud pulses by blue diamonds (*Courtesy* Vaisala Inc.)

morning and early afternoon. The 7-h period from midday at 1200–1900 MST (12:00–7:00 p.m.) accounts for 70% of the day's lightning over the state. This diurnal curve is remarkably similar to the time of day shown for the entire U.S. detected by the NLDN (Holle 2014). Comparisons with the U.S. curve are given as follows:

- The minimum is reached from 08:00 to 10:00 local time in both the U.S. and Arizona.
- The maximum is reached from 16:00 to 18:00 local time for the U.S., while Arizona lightning starts to trend downward slightly after 15:00 MST (3:00 p.m.).
- The U.S. curve has a much slower decrease after 18:00 local time than Arizona where lightning starts to trend downward quite quickly around that time.
- The reason for the delayed reduction in late afternoon and evening in the U.S. is most often due to thunderstorms that linger and grow larger into

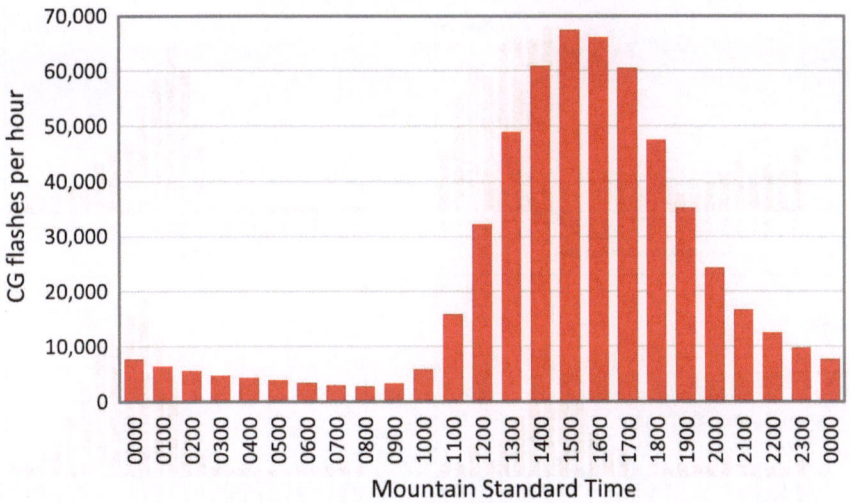

Fig. 4.20 CG flash counts by hour per year over Arizona from 2009 to 2018 (*Courtesy Vaisala Inc.*)

the evening and nighttime hours on the Central Plains during the summer (Holle 2014).

- Watson et al. (1994b) have similar diurnal results but in a different format for Arizona.

Hourly city-by-city changes are shown in Fig. 4.21 for the major cities within Arizona shown in Figs. 4.7, 4.8, 4.9, and 4.10. The following are the core results by hour:

- Flagstaff and Tucson have nearly the same features as for the entire state in Fig. 4.20. That is, a rapid increase from 1000 MST (10:00 a.m.) to a peak in mid-afternoon. The decrease is slower in the evening in Tucson than Flagstaff due to thunderstorms reaching Tucson from the east and south near and after sunset.
- Phoenix lightning is much less frequent compared with Flagstaff and Tucson. Lightning activity peaks in mid-afternoon but continues well after dark due to outflow-generated thunderstorms arriving from distant mountains to the northeast through southeast. Storms sometimes occur through the night and linger into the morning hours; some of this morning lightning is due to decaying tropical systems in September and October.
- Yuma has less lightning than Phoenix and activity is concentrated during a few hours from 1500 to 1900 MST (3:00–7:00 p.m.). Minimal lightning occurs at all other hours.

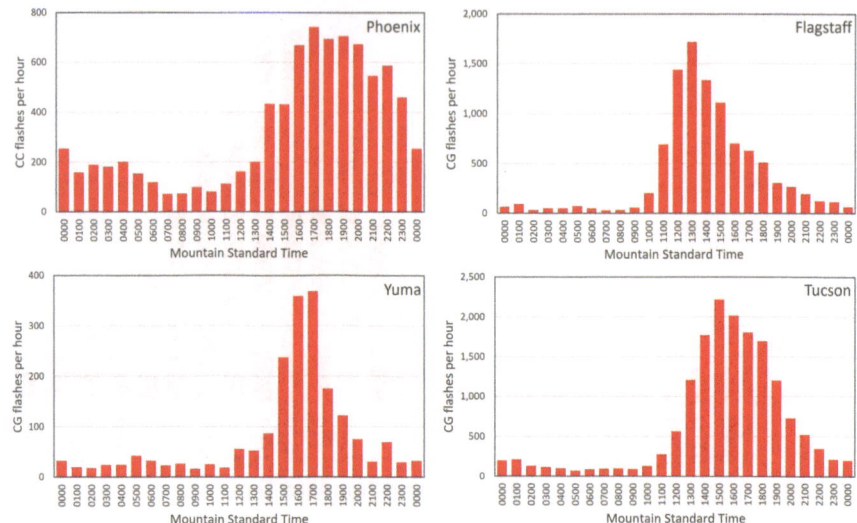

Fig. 4.21 CG flash counts by hour per year for four cities in Arizona from 2009 to 2018 (*Courtesy* Vaisala Inc.)

4.4   Special Topics in Lightning Occurrence

4.4.1   Grand Canyon

There are several reasons to pay special attention to lightning in the Grand Canyon. First, it has a remarkable pattern as shown in Fig. 4.22. This unusual distribution of lightning is not replicated elsewhere due to the depth of the canyon as well as its large size as the Colorado River flows a long distance from east to west. Second, it is the location of more lightning deaths and injuries than anywhere else in Arizona (Sect. 5.1.7). Third, this pattern is rather weird! Dr. Ken Cummins in the Department of Hydrology and Atmospheric Sciences at the University of Arizona developed this view in Fig. 4.22 that plots every CG stroke that was detected by the NDLN from 2008 through 2017; an update is in Holle et al. (2020). Subsequently, he and others considered various hypotheses and made plans for a field program to the canyon described in Sect. 7.2.2 and in Cummins et al. (2019).

An additional exploration of this pattern is in a journal article co-authored by Dr. Cummins on lightning in U.S. National Parks wherein the Grand Canyon is one of the six featured parks due to the large number of visitors (Holle et al. 2020). A more complete context for Fig. 4.22 is provided in Fig. 4.23. First, note the extreme topography in panel (a), where the deeply incised canyon cuts through the high-elevation plateaus located to

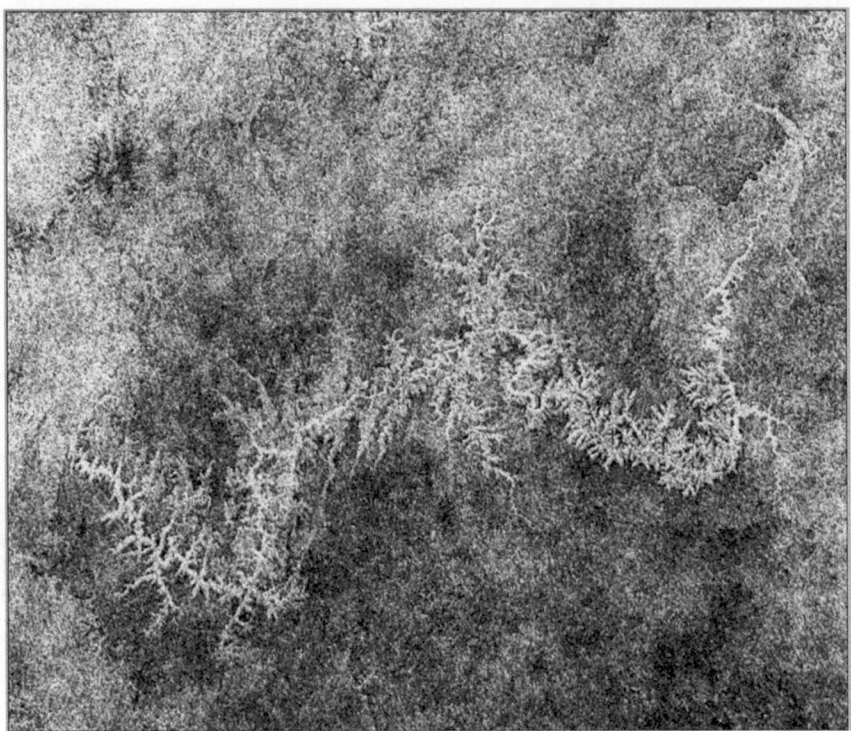

Fig. 4.22 Individual CG strokes detected by the NLDN in the vicinity of the Grand Canyon from 2008 to 2017. © American Meteorological Society. Used with permission

the north and south. The flash density (b) is greater away from the rim on both sides, while the dendritic pattern of the individual strokes (c) is most prominent within the park boundary. Familiar landmarks to visitors are in Fig. 4.23d. A fly-by version of Fig. 4.23 is available in Supplementary Materials to Holle et al. (2020) at https://journals.ametsoc.org/view/journals/wcas/13/3/WCAS-D-19-0155.1.xml. These fly-by views show the extent of the deficit in lightning within the canyon while CG flashes are greater on the rims and other promontories. The research questions raised here are explored more extensively in Sect. 7.2.2.

Grand Canyon Music

The Grand Canyon Suite was composed from 1929 to 1931 by the American musician Ferde Grofe. Movement V titled "Cloudburst" has a wonderful evocation of a thunderstorm over the Canyon. More recently, Nicholas Gunn

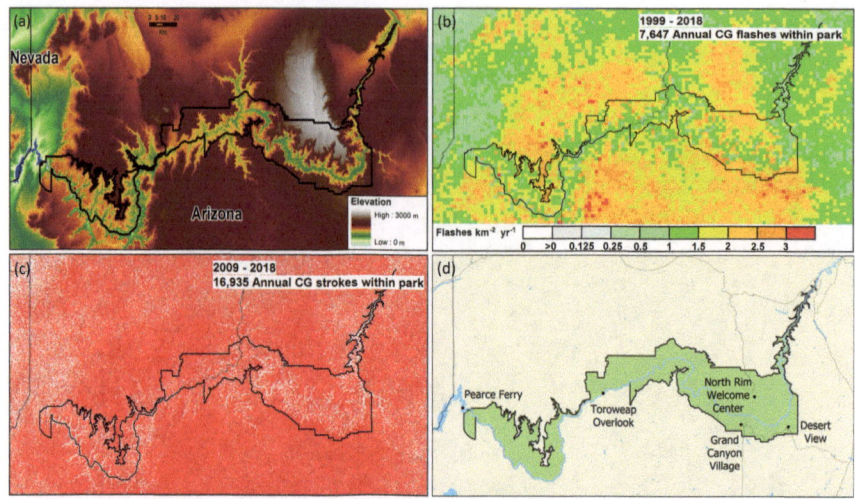

Fig. 4.23 a Grand Canyon National Park elevation map, **b** 1999–2018 CG flash density on a grid of 2 km by 2 km, **c** point plot of 2009–2018 CG strokes, and **d** significant features and locations within the park. © American Meteorological Society. Used with permission

composed "The Music of The Grand Canyon" that illustrates the continuing impact of the canyon on people!

4.4.2   Southeast Mountains

If there is only a small amount of lightning on a monsoon-season afternoon in Arizona, it is likely to be in the southeast mountains and to the north along the eastern border with New Mexico. Storms can form as early as 1000 MST (10:00 a.m.) on the slopes of the mountains, so hikers need to be aware of this early start (Sect. 5.2). On some afternoons in the monsoon months, thunderstorms will organize across all southeast Arizona to form a broad area of lightning moving steadily to the northwest. These storms may reach Tucson and contribute to the late afternoon peak in Fig. 4.21. On some days, they will continue to generate new thunderstorms on the forward edge of the advancing boundary and move past Tucson to the northwest. Then, they may produce dust storms (haboobs) over the lower desert northwest of Tucson, and several times a year they will result in the nighttime lightning maximum when they reach the Phoenix area (Fig. 4.21).

4.4.3 Mogollon Rim

Similar to the southeast mountains, Mogollon Rim storms are mostly during afternoons in the monsoon months. Also, lightning can occur as early as 1000 MST (10:00 a.m.) on the slopes of the escarpment dividing much of southern from northern Arizona (Fig. 2.1). Since this region has a very large number of recreational activities in summer, early storms can be problematic for lightning safety. Figure 4.3 shows a lightning maximum extending from the White Mountains in Greenlee County to the west-northwest. It is the slope of the terrain here that enhances lightning occurrence, not the absolute elevation. To the south of the Mogollon Rim, lightning in the lower desert is less frequent. To the north is a high plateau where lightning is also less common due to a lack of large elevation changes.

A particular type of lightning event can develop on monsoon afternoons called a "Rim Shot" when thunderstorms form along the elevation change and under the correct conditions move steadily to the southwest. Rim-generated storms can reach as far as Phoenix and result in the evening and nighttime maximum as shown in Fig. 4.21. This situation was identified from a meteorological point of view by Watson et al. (1994b).

On a few days each summer, thunderstorms from both the southeast mountains and Mogollon Rim will reach the lower deserts, including the Phoenix metropolitan area. The resulting convergence of the two areas of thunderstorms can produce some of the most significant lightning events each year in central Arizona.

4.4.4 Western Desert

Lightning is very infrequent in the lower deserts of the western part of the state. What is missing? There are no large-scale elevation changes; instead, there is a slow descent from east to west toward the Colorado River that is near sea level and forms the western border with California and Nevada. Storms originating over the southeast mountains and Mogollon Rim are usually not able to descend this long distance and maintain intensity. Instead, lightning in this region is partially due to remnants of tropical systems, as represented by the September peak at Yuma in Fig. 4.18. Up to half of the annual rain and likely also half of the annual lightning in this region is from such systems (Reyes and Mejia Trejo 1991; Corbosiero et al. 2009; Mejia et al. 2015). The rest of the western desert lightning tends to be generated by thunderstorms later in the monsoon season when there is deeper lower and middle level moisture. Sometimes, storms that form around Tucson and the nearby international border can maintain themselves all the way to Yuma during the evening and nighttime hours when the upper air flow is deep and from the east.

References

Adams DK, Comrie AC (1997) The North American monsoon. Bull Amer Meteor Soc 78:2197–2213

Adams D, Minjarez-Sosa C, Serra Y et al (2014) Mexican GPS tracks convection from North American Monsoon. EOS, Trans Amer Geophys Union 95. https://doi.org/10.1002/2014EO070001

Albrecht R, Goodman SJ, Buechler DE et al (2016) Where are the lightning hotspots on earth? Bull Amer Meteor Soc 97:2051–2068

Corbosiero KL, Dickinson MJ, Bosart LF (2009) The contribution of Eastern North Pacific tropical cyclones to the rainfall climatology of the Southwest United States. Mon Wea Rev 137:2415–2435

Cummins KL, Brooks WA, Holle RL (2019) Mapping the impact of local terrain on lightning ground attachment location. Paper presented at annual meeting of American geophysical union. https://agu.confex.com/agu/fm19/meetingapp.cgi/Paper/516604

Douglas MW, Maddox RA, Howard K (1993) The Mexican monsoon. J Climate 6:1665–1677

Farfán LM, Zehnder JA (1994) Moving and stationary mesoscale convective systems over Northwest Mexico during the Southwest Area Monsoon Project. Weather Forecast 9:630–639

Higgins W, Gochis D (2007) Synthesis of results from the North American Monsoon Experiment (NAME) process study. J Clim 20:1601–1607

Higgins RW, Shi W (2005) Relationships between Gulf of California moisture surges and tropical cyclones in the eastern Pacific Basin. J Clim 18:4601–4620

Holle RL (2014) Diurnal variations of NLDN-reported cloud-to-ground lightning in the United States. Mon Wea Rev 142:1037–1052

Holle RL, Murphy MJ (2015) Lightning in the North American Monsoon: an exploratory climatology. Mon Wea Rev 143:1970–1977

Holle RL, Brooks WA, Cummins KL (2020) Lightning occurrence and casualties in U.S National Parks. Wea Clim Soc 12:525–540

Holle RL, Dewan A, Said R et al (2019) Fatalities related to lightning occurrence and agriculture in Bangladesh. Int J Disaster Risk Reduction 41:15

Holle RL, Selover N, Cerveny R et al (2015) The weather and climate of Arizona. Weatherwise 68:13–19

Jana S, Rajagopalan B, Alexander MA et al (2018) Understanding the dominant sources and tracks of moisture for summer rainfall in the southwest United States. J Geophys Res Atmos 123:4850–4870

Medici G, Cummins KL, Cecil DJ et al (2017) The intracloud lightning fraction in the contiguous United States. Mon Wea Rev 145:4481–4499

Mejia JF, Douglas MW, Lamb PJ (2015) Observational investigation of relationships between moisture surges and mesoscale- to large-scale convection during the North American monsoon. Int J Climatol 36:2555–2569

Murphy MJ, Cramer JA, Said RK (2021) Recent history of upgrades to the U.S. National Lightning Detection Network. J Atmos Oceanic Tech 38:573–585

Reyes S, Mejia Trejo A (1991) Tropical perturbations in the eastern Pacific and the precipitation field over northwestern Mexico in relation to the ENSO phenomenon. Int J Climatol 11:515–528

Said R (2017) Towards a global lightning locating system. Weather 72:36–40

Said R, Inan U, Cummins KL (2010) Long-range lightning geolocation using a VLF radio atmospheric waveform bank. J Geophys Res 115:D23108. https://doi.org/10.1029/2010JD013863

Valine WC, Krider EP (2002) Statistics and characteristics of cloud-to-ground lightning with multiple ground contacts. J Geophys Res 107. https://doi.org/10.1029/2001JD001360

Watson AI, Holle RL, López RE (1994a) Cloud-to-ground lightning and upper-air patterns during bursts and breaks in the southwest monsoon. Mon Wea Rev 122:1726–1739

Watson AI, López RE, Holle RL (1994b) Diurnal lightning patterns in Arizona during the southwest monsoon. Mon Wea Rev 122:1716–1725

Zhang D, Cummins KL, Bitzer P et al (2019) Evaluation of the performance characteristics of the lightning imaging sensor. J Atmos Oceanic Tech 36:1015–1031

Zipser EJ, Cecil DJ, Liu C et al (2006) Where are the most intense thunderstorms on earth? Bull Amer Meteor Soc 87:1058–1071

5

Human Impacts, Damages, and Benefits from Lightning in Arizona

Abstract Lightning has impacted people and damaged property in Arizona in some unexpected ways. This chapter presents the first history of known lightning deaths and injuries for nearly a century back to 1926. The time of year and day, age, gender, county, activity, and location of people affected by lightning are also described in detail as found mainly in the National Weather Service publication *Storm Data*. Lightning safety advice is provided that is focused on the situation in Arizona and based on the policies recommended by specialists in this topic in the U.S. and elsewhere. There are only two common reliably lightning-safe places, large well-constructed buildings, and fully enclosed metal-topped vehicles, unless lightning-specific protection has been installed. Lightning damages in Arizona are also depicted in detail by time and location, including impacts on homes and other property, power utilities, insurance, agricultural, and forestry losses. Some beneficial features of lightning are included to complete the chapter.

5.1 Lightning Deaths and Injuries in Arizona

Here are the long-term and short-term totals of lightning fatalities and injuries in Arizona, along with their rankings among the U.S. states:

- During the 63 years from 1959 through 2021, 84 people have been killed by lightning in Arizona, which ranks 23rd among the 50 U.S. states. The rate is 0.42 deaths per million people per year from 1959 to 2021, which is also ranked 23rd.

© Springer Nature Switzerland AG 2023
R. L. Holle and D. Zhang, *Flashes of Brilliance*,
https://doi.org/10.1007/978-3-031-19879-3_5

- In the decade from 2010 to 2019, 13 people were killed in Arizona, ranked 4th in the U.S. The rate during this period is 0.19 deaths per million people per year, which is ranked 8th.
- There were no lightning deaths in 2020 or 2021.
- About ten times as many people are estimated to be injured as are killed by lightning in the U.S. (Cherington et al. 1999). Using this ratio, an estimated 840 people were injured in the 63-year period from 1959 through 2021, and 130 people were injured in the decade from 2010 to 2019.
- Underreporting of lightning deaths, and especially injuries, is known to have occurred, especially in earlier years (López et al. 1993).

How do we know so much about these casualties? The National Weather Service began collecting data on all types of weather impacts starting in 1959 in a publication aptly named *Storm Data*. The monthly issues of *Storm Data* summarize all significant weather events affecting people as well as weather events causing major damage from hurricanes, tornadoes, strong winds, hail, snow, ice, heat, cold, and of course lightning, in a standardized method. A description of how U.S. lightning fatalities and injuries were extracted from *Storm Data* in the earlier years is in Curran et al. (2000). Each of the 122 National Weather Service Forecast Offices across the U.S. has an employee whose duties include gathering all qualifying events within the office's area of responsibility at the end of each month. Until around 2000, the reports were collected primarily by reviewing reports supplied by newspaper clipping services. Since then, the reports have transitioned to arrive from city and county emergency managers, police and sheriff offices, online social media reports, and broadcast media. The reports are then sent to the national collection office at the National Centers for Environmental Information (NCEI) of the National Oceanic and Atmospheric Administration (NOAA). At the end of the year, a summary is made of each type of weather by state. These permanent records of significant weather events may be used in insurance and other investigations. It also forms the basis for research and planning by greatly furthering the understanding of when and where various types of important weather events tend to have occurred. Other countries may have more frequent winter storms or flooding problems or other types of recurring weather impacts, but the U.S. has every weather type somewhere in the country every year!

Storm Data

The publication *Storm Data* that is available online at https://www.ncdc.noaa. gov/stormevents/ has a wealth of choices of ways to look at the data.

Secondary weather-related deaths and injuries are usually not included in *Storm Data*. This policy affects lightning most often when a house is set on fire and a person is killed or injured. The reason for this policy has been developed in response to the difficulty of reporting vehicle crashes in winter, where weather is a factor but difficult to disconnect from other factors. As a result, secondary lightning deaths and injuries are rarely included in *Storm Data*, and only a few of these cases have been included in recent years.

Notice the change in the numbers and rates between the 61-year and the decadal periods. Why are they so different? What effect has the population growth of Arizona had on this record, since Arizona grew from 1,261,000 people in 1959 to 7,276,000 in 2021? Is there a certain type of scenario that accounts for the number of people killed and injured by lightning? Were there different types of events in the 1960s and 1970s compared to now? These factors will be explored in the following sections.

5.1.1 Annual Fatality Totals, 1926–2021

We now present the first long-period record of lightning deaths for Arizona. There are two datasets, one before and one after *Storm Data* began to be compiled in 1959.

From 1926 through 1958, Table 5.1 lists deaths by year; no annual injury data are available. There is one year (1926) with nine deaths, then the number steadily decreases. The source of this information is unusual. Dr. E. Philip Krider of the University of Arizona (UArizona) provided the authors with unique resources from his library. In fact, he had kept the early data that were collected in 1959 and 1960 by Dr. James E. McDonald, a faculty member of the Dept. of Atmospheric Sciences at UArizona at that time. Two sources were used to develop handwritten lists compiled by Dr. McDonald; both sources mentioned their difficulty in collecting lightning fatality data at their respective offices:

- 1926–1936 and 1950–1958: Provided to Dr. McDonald in a letter dated 8 September 1959 from the Arizona State Dept. of Health in Phoenix.

Table 5.1 Number of lightning-related deaths in Arizona from 1926 to 1958 based on statistics collected from Arizona and U.S. health agencies by Dr. James E. McDonald of the UArizona

Year	Lightning deaths	Year	Lightning deaths
1926	9	1940	1
1927	7	1941	3
1928	6	1942	5
1929	7	1943	6
Sum	29	1944	1
1930	6	1945	2
1931	3	1946	0
1932	6	1947	4
1933	3	1948	1
1934	6	1949	1
1935	3	Sum	24
1936	5	1950	2
1937	3	1951	5
1938	2	1952	4
1939	1	1953	1
Sum	38	1954	2
		1955	1
		1956	2
		1957	2
		1958	0
		Sum	19

- 1937–1957: Provided to Dr. McDonald in a letter dated 13 July 1960 from the U.S. Dept. of Health, Education, and Welfare, Public Health Service, National Office of Vital Statistics, Washington, 25, D.C.

Starting in 1959, *Storm Data* has both death and injury data as shown in Table 5.2. Almost all entries are from the online *Storm Data*. From 1959 through 1995, monthly reports have not been scanned so they are not in an online database. As a result, it was necessary to search manually through each monthly report during these years for lightning in Arizona. Starting in 1996, a county and month can be entered by the user in the *Storm Data* website to obtain reports of only lightning impacts. In recent years, some additional lightning injury cases were found in newspapers and web articles. In two cases, deaths from house fires that were classified as indirect have been included.

Table 5.2 Number of lightning-related deaths, injuries, and events in Arizona from 1959 through 2021 based on *Storm Data* and additional newspaper articles and web resources

Year	Deaths	Injuries	Events	Year	Deaths	Injuries	Events
1959	1	2	3	1990	0	2	2
Sum	*1*	*2*	*3*	1991	4	12	4
1960	0	2	1	1992	2	6	3
1961	1	1	1	1993	1	0	1
1962	3	0	3	1994	2	6	3
1963	2	1	3	1995	1	17	5
1964	0	1	1	1996	0	7	5
1965	2	5	5	1997	2	6	7
1966	8	5	6	1998	1	3	1
1967	1	3	3	1999	1	13	12
1968	1	0	1	*Sum*	*14*	*72*	*43*
1969	1	2	2	2000	2	27	11
Sum	*19*	*20*	*26*	2001	0	2	2
1970	0	1	1	2002	4	8	1
1971	0	3	2	2003	0	8	3
1972	0	0	0	2004	0	5	2
1973	3	9	6	2005	1	2	2
1974	1	2	2	2006	3	4	4
1975	0	0	0	2007	1	9	3
1976	1	6	4	2008	0	6	4
1977	1	0	1	2009	0	2	2
1978	2	1	3	*Sum*	*11*	*73*	*34*
1979	0	4	3	2010	1	2	2
Sum	*8*	*25*	*22*	2011	2	11	7
1980	0	1	1	2012	0	0	0
1981	1	5	4	2013	4	7	6
1982	3	0	3	2014	2	1	3
1983	1	4	2	2015	3	9	5
1984	5	3	4	2016	1	3	2
1985	0	0	0	2017	0	7	1
1986	2	2	2	2018	1	12	5
1987	2	1	3	2019	0	1	1
1988	1	5	4	*Sum*	*14*	*53*	*32*
1989	0	1	1	2020	0	0	0
Sum	*15*	*22*	*24*	2021	0	2	1
				Sum	*0*	*2*	*1*

5.1.2   Decadal Fatality Rates

By decade, Table 5.3 and Fig. 5.1 show the rapid decline in rates of lightning deaths per million people from the late 1920s to the 2020s. The rate
is extraordinarily high in the early years and quickly drops. For comparison
around the world, Holle (2016) categorized recent death rates on a national
basis into three groups:

Table 5.3 Decadal totals of the number of lightning-related deaths, injuries and
events, and population-weighted fatality rates per million per year in Arizona from
1926 through 2021 based on Tables 5.1 and 5.2

Period	Deaths	Injuries	Events	Deaths per million people per year
1926–1929	29	xx	xx	17.2
1930–1939	38	xx	xx	8.8
1940–1949	24	xx	xx	4.0
1950–1959	20	xx	3	1.9
1960–1969	19	20	26	1.2
1970–1979	8	25	22	0.4
1980–1989	15	22	24	0.5
1990–1999	14	72	43	0.3
2000–2009	11	73	34	0.2
2010–2019	14	53	32	0.2
2020–2021	0	2	1	0
Sum	*192*	*267*	*185*	

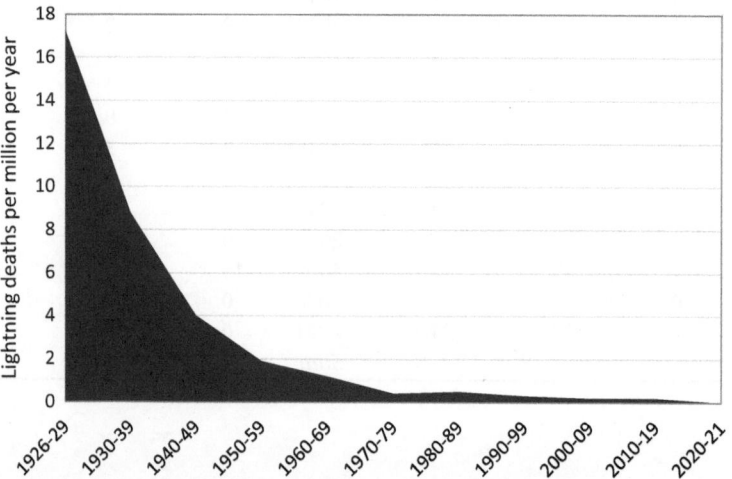

Fig. 5.1 Decadal Arizona lightning deaths per million people per year from 1926
through 2021

1. The highest rate is over 5.0 deaths per million people per year, seen only at the present time in Africa and perhaps some Asian countries.
2. An intermediate range of 0.5–5.0 deaths per million per year applies in many developing nations.
3. The lowest rate of under 0.5 deaths per million per year applies to the developed nations of the U.S., Canada, western Europe, Australia, New Zealand, and Japan.

In other words, the lightning fatality rate weighted by population in Arizona in the last century went from what is now considered applying only to developing nations to a rate that is typical of very developed nations. Reasons will be identified later in Sect. 5.1.8.

Despite these seemingly low rates, let's not underestimate the impact of any lightning death or injury. In the last decade, between one and two people are killed by lightning per year in Arizona. The known reports show about five injuries per year; however, Cherington et al. (1999) found in an intensive search of local medical records in Colorado that as many as ten times as many people are injured as are killed by lightning when all data are examined. Let's make a few assumptions to indicate that there is a 1 in 1,250 probability that a person in Arizona will be killed or injured by lightning or be a close relative or friend of a victim at today's casualty rate, over an 80-year life span. This is explored more completely in Sect. 5.2.1.

5.1.3 Month and Hour

The timing is now shown within the years for events since 1959 in the previous tables. Table 5.4 indicates a strong concentration of deaths and injuries in July and August, comprising nearly two thirds of the year's total. September follows in frequency, and June has somewhat fewer casualties than September. The rest of the months contribute minimally to the annual total of lightning casualties.

By the time of day, Table 5.5 shows information for when it is reported from various sources. A number of reports indicate only the general time such as afternoon, while others are more exact. When "afternoon" events are combined with those directly reported as occurring between 12:00 and 6:00 p.m., 60% of the events are found to be during this period.

Correlations of cloud-to-ground (CG) flashes with the month and time of day of casualty events are shown in Fig. 5.2. The correlation is very high on the monthly scale ($r = 0.9856$), while the correlation is also very good for time of day ($r = 0.8639$).

Table 5.4 Month of 185 lightning-related casualty events, and percent of annual total across Arizona from 1959 to 2021

Month	Events per month	Percent of annual total
January	1	0.5
February	3	1.6
March	1	0.5
April	1	0.5
May	10	5.4
June	16	8.6
July	70	37.8
August	50	27.0
September	24	13.0
October	7	3.8
November	1	0.5
December	1	0.5

Table 5.5 Time of day in Mountain Standard Time of 151 lightning-related casualty events with known times, and percent of total within Arizona from 1959 to 2021

Time	Events	Percent
General		
Morning	2	1.3
Afternoon	15	9.9
Evening	9	6.0
Specific		
Midnight to 0600 (midnight–6 a.m.)	3	2.0
0600–1200 (6 a.m.–noon)	8	5.3
1200–1500 (noon–3 p.m.)	37	24.5
1500–1800 (3–6 p.m.)	39	25.8
1800–2100 (6–9 p.m.)	26	17.2
2100–Midnight (9 p.m.–midnight)	12	7.9

5.1.4   Gender and Age

Who were the people killed and injured? Of those 210 people whose gender information was reported, Table 5.6 shows the gender to be most often male. In fact, it is overwhelmingly so, as 82% of the casualties are male. In terms of age when it was reported (Table 5.7), nearly one third (29.5%) are between the ages of 10 and 19. The next largest groups are 20–29 and 30–39 years old.

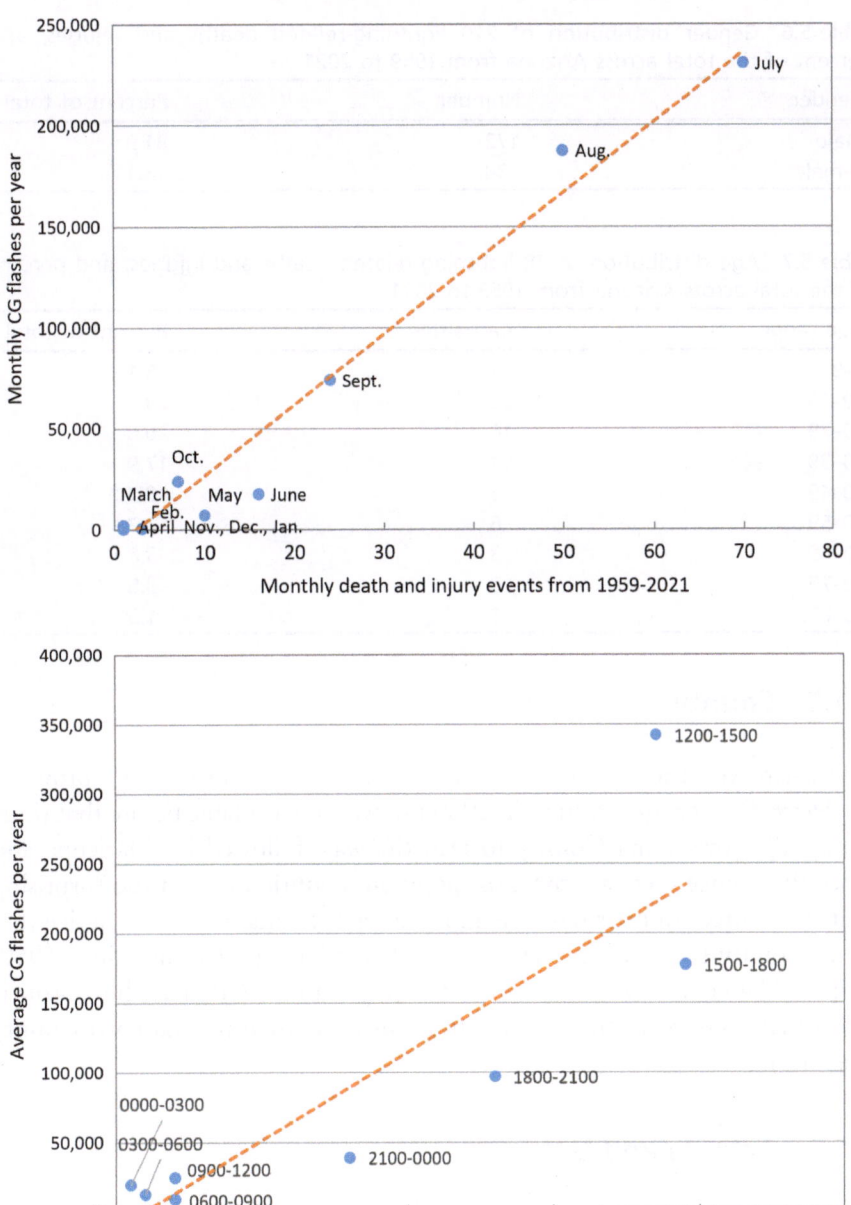

Fig. 5.2 Top panel: Correlation of monthly death and injury events with monthly CG flashes per year. Bottom panel: Correlation of 3-hourly death and injury events with 3-hourly CG flashes per year

Table 5.6 Gender distribution of 210 lightning-related deaths and injuries, and percent of the total across Arizona from 1959 to 2021

Gender	Number	Percent of total
Male	172	81.9
Female	38	18.1

Table 5.7 Age distribution of 78 lightning-related deaths and injuries, and percent of the total across Arizona from 1959 to 2021

Age range	Casualties	Percent of total
0–9	4	5.1
10–19	23	29.5
20–29	16	20.5
30–39	14	17.9
40–49	8	10.3
50–59	6	7.7
60–69	3	3.8
70–79	3	3.8
80–89	1	1.3

5.1.5   County

And where were they located within Arizona? County information is provided for *Storm Data* entries starting in 1959 but was not available before that time. Table 5.8 shows Pima County to lead the way, followed by Maricopa and Coconino. Since these are the most populous counties, it's not too surprising that these three counties have the most casualty events. However, notice that Pima County has more casualty events than Maricopa County since 1959, despite Maricopa having five times the population of Pima. The reason is likely that there is much more lightning in Pima than in Maricopa County (Fig. 4.3).

5.1.6   Activity and Location

What were people doing and where were they located when they were killed or injured by lightning in Arizona between 1959 and 2021? The next series of tables takes the information in *Storm Data* and provides general, then increasingly specific insights.

Table 5.8 Counties where 185 lightning-related deaths and injuries events occurred, percent of total, and rank within Arizona from 1959 to 2021. La Paz county was established in 1983 from the northern portion of Yuma county

County	Number	Percent	Rank
Pima	49	26.5	1
Maricopa	38	20.5	2
Coconino	34	18.4	3
Cochise	16	8.6	4
Apache	13	7.0	5
Gila	6	3.2	6
Pinal	6	3.2	6
Graham	4	2.2	8
Yavapai	4	2.2	8
Greeenlee	3	1.6	10
Mohave	3	1.6	11
Navajo	3	1.6	11
Santa Cruz	3	1.6	11
Yuma	3	1.6	11
La Paz [since 1983]	0	0	15

Table 5.9 Activity and/or location overview of 151 lightning-related death and injury events within Arizona from 1959 to 2021

Activity/location	Events
Recreation (Table 5.10)	21
Vehicles (Table 5.11)	19
Occupation (Table 5.12)	18
Home/neighborhood (Table 5.13)	17
Under or near trees	13
Grand Canyon (Sect. 5.1.7)	12
Hiking/climbing	9
Border crossers	6
Boat	5
Military	3
Open field	2
Miscellaneous (Table 5.14)	26

In the most general grouping in Table 5.9, recreation is at the top, followed by vehicles, working in an occupation, then in the home or neighborhood. Each of these is expanded later in Tables 5.10, 5.11, 5.12, 5.13, 5.14. Comments on the other categories in Table 5.9 and not in other tables are the following:

Table 5.10 Activity and/or location during 21 lightning-related recreation death and injury events within Arizona from 1959 to 2021

Activity/location	Events
Fishing	5
Golf	4
Baseball	3
Camping	3
Swimming	2
Horse riding or walking	2
Hunting	1
Athletic field	1

Table 5.11 Activity and/or location during 19 vehicle lightning-related death and injury events within Arizona from 1959 to 2021

Activity/location	Events
Motorcycle	5
Inside car/truck	4
Entering car	2
Car dealership	1
Driveway	1
Gas station	1
Helping motorist	1
Highway flagger	1
Near van	1
Parking lot	1
Truck bed	1

Table 5.12 Activity and/or location during 18 lightning-related occupational death and injury events within Arizona from 1959 to 2021

Activity/location	Events
Field workers (all 1960s + 1970s)	6
Utility workers	4
Firefighters	3
Removing ornaments from tall tree	1
Herding sheep	1
Mining camp	1
School maintenance	1
Delivering newspapers	1

- Under or near trees occurs in 13 events with any type of casualty; these are typically in more than one category.
- Grand Canyon is the location of 12 events and will be considered later in Sect. 5.1.7.

Table 5.13 Activity and/or location during 17 lightning-related home and neighborhood death and injury events within Arizona from 1959 to 2021

Activity/location	Events
Telephone	5
Inside house	4
Fire [one house, one mobile home]	2
Roof	2
Kitchen	1
Neighborhod	1
Sprinkler system repair in yard	1
Yard	1

Table 5.14 Activity and/or location during 26 lightning-related miscellaneous death and injury events within Arizona from 1959 to 2021

Activity/location	Events
Walking	5
Near building	3
Construction	3
Fence	2
Airport	2
Archeology transcept	1
Operating amateur radio	1
Grave	1
Ladder	1
Park	1
Near dam	1
Playing	1
Running for shelter	1
Scenic overview	1
Watching flood	1
Under umbrella	1

- Hiking and climbing on trails and mountains is the activity/location of nine events.
- Border crossers (illegal immigrants) are the source of six events, primarily in recent years, often with multiple casualties and all under trees or in open areas.
- Boats account for 5 events, military for 3 events, 2 were in open fields, and 26 events are not readily categorized, so they are listed as Miscellaneous in Table 5.9.

In the recreation category (Table 5.10), golf had often been perceived as the number one source of lightning casualties in prior decades. However, the attention given to golf course safety has greatly reduced this former high ranking to only 4 of the 151 events. Instead, fishing has come in first in Arizona and matches the large proportion that it holds nationally at the National Lightning Safety Council (www.lightningsafetycouncil.org) and National Weather Service https://www.weather.gov/safety/lightning-safety websites. Baseball events occurred in the earlier years of *Storm Data*, but lightning safety advice has also emphasized this vulnerability to group sports (Walsh et al. 2013). Camping has three events; it should be recognized that tents never provide protection against lightning (Sect. 5.2).

In the vehicle category (Table 5.11), motorcycles are the most common; a previous review of motorcycle lightning-related incidents for all areas of the world was made by Cooper and Holle (2007). The next category with four events of being inside a car or truck involves injuries but no fatalities. The injuries are typically minor such as numbness, dizziness, or tingling sensations, along with some events consisting of flying debris from trees piercing the vehicle. Being inside a fully enclosed metal-topped vehicle is very safe from lightning but can be scary. Nevertheless, it is an excellent source of safety and immensely preferable to being outside the vehicle when a large substantial building is not readily available (see story in the paragraph titled "*Flying Through the Air*" below). In contrast, the two cases of entering a car are highly vulnerable. In this situation, a person has step potential (Sect. 5.2.2), when one foot is on the ground and an arm or foot is touching the car when lightning strikes the car or the ground nearby; these situations outside the protection of the Faraday-cage-like vehicle situation are very dangerous. All of the additional single events related to vehicles are in the open, outside the protection of a fully enclosed metal-topped vehicle, so they are very dangerous (Holle 2008b).

Flying Through the Air

The first author of this book has attended several annual meetings of the Lightning Strike and Electric Shock Survivors International (LSESSI), a non-profit organization dedicated to survivors, their families, and other interested parties (https://www.lightning-strike.org). At a May meeting a few years ago, a young man from France told his story. At a beach on the Mediterranean Sea the previous summer, he and his friends decided to go to their car when lightning was approaching. Everyone except the man went inside the car, then lightning hit the car or very nearby. All inside the car were safe and uninjured, but they watched in amazement as he flew through the air. At the LSESSI meeting a year later, he was clearly not doing well due to complex neurologic issues. The difference is evident between being inside a fully enclosed metal-topped vehicle, and being anywhere outside a lightning-safe vehicle or building.

In the occupation category (Table 5.12), the most common one was field workers. These events in the 1960 and 1970s were primarily in Pima County but are no longer a factor due to massive agricultural mechanization. The U.S. has shifted from an 80% rural population in the early twentieth century to less than 20% in recent years due to changing farming practices such that labor-intensive agricultural is no longer a source of frequent lightning casualties. One can speculate that these large numbers of fatalities in Arizona in the 1920s and 1930s (Table 5.1) may have been agricultural, but the data sources do not provide this information. In addition, the cases in Table 5.12 from the 1960s and 1970s involved more than one injury or death to people working in the fields.

Utility workers and firefighters in the occupation category of Table 5.12 have several events, all involving being outdoors, sometimes in thunderstorms. In both categories, going inside a fully enclosed metal-topped vehicle may have provided a ready refuge from lightning.

In the home and neighborhood category (Table 5.13), telephones are the most common source of injury from lightning. These events can only occur when a wired landline is involved, so this risk is mostly gone with present-day cordless and portable cell phones. There can be no injury from lightning when a phone is not connected to a line that conducts the lightning current into a house from a nearby utility pole. This statement does not apply outside a lightning-safe house or building—if a person is outside in the open, under a tree, or otherwise exposed to lightning, a person is equally vulnerable to lightning whether they are holding any type of phone or not.

Question

Does it make any sense to say that cell phones by themselves can attract lightning as is sometimes stated? Why not?

The next largest category in Table 5.13 is inside a house. Lightning injuries inside homes are not rare but are usually not too serious although they can be debilitating in some cases. Nearly all U.S. lightning deaths inside houses occur to people who are unable to escape a house set on fire at night, and are elderly or very young, and may be physically or mentally disabled (Holle 2010). The house is a very safe place from lightning but injuries should be avoided by not being in contact with plumbing and wiring such as in the kitchen or bathroom, and by not being in direct contact with wired computer connections.

In the miscellaneous category (Table 5.14), walking is the most common source of injury from lightning; many of these are in open areas. There are three lightning injury and death events stated as near buildings, which indicates that the person was outside the protection provided by being inside a large substantial frequently occupied building or a fully enclosed metal-topped vehicle. Construction can also be a source of lightning death and injury. For example, a worker is vulnerable if a house is being built but the wiring and plumbing are not complete enough to provide protection if the home is struck. Many other miscellaneous events listed in Table 5.14 are from the descriptions in *Storm Data* since 1959.

It's worth looking at the events with larger numbers of casualties, as in Table 5.15. Two categories are noted, at the Grand Canyon (Sect. 5.1.7) and during hiking and climbing. The largest single fatality event involved four border crossers killed and eight injured when they, unfortunately, sought lightning protection under a tree.

Table 5.15 Ten events with five or more lightning-related casualties within Arizona from 1959 to 2021

Year	Number	County
Grand Canyon		
2000	7 injured	Coconono
2011	1 killed, 5 injured	Coconino
Hiking/climbing		
1992	5 injured	Pima
2015	1 killed, 7 injured	Coconino
Other		
1973	1 killed, 4 injured	Mohave
	Baseball	
1995	10 injured	Coconino
	Unknown	
2002	4 killed, 8 injured	Cochise
	Border crossers under tree	
2003	5 injured	Pima
	Cable workers	
2007	5 injured	Pima
	Archaeleogy transcept	
2017	7 injured	Coconino
	Military; working on helicopter	

5.1.7 Grand Canyon

A recent journal article addressed lightning in all mainland U.S. national parks, including Grand Canyon National Park (Holle et al. 2021). The locations of 14 events within the park in Fig. 5.3 indicate a large cluster near the Visitor Center and Grand Canyon Village on the South Rim. In terms of CG flash density, lightning occurrence over the entire Grand Canyon National Park is ranked 19th among the 45 mainland parks. The difficulty with respect to lightning occurs because it was the second most-visited national park by 6,380,495 people in 2018. The flash rate itself is not large, but it has more lightning casualty events than all but a few parks due to the large number of visitors. In addition, the assembly of so many people on the rims is a dangerous situation for reasons given in more detail in Sect. 7.2.2.

Additional information on the events with known data are the following for the 13 incidents prior to 2021 involving injury or death in Holle et al. (2021) plus one event in 2021:

- All casualty events are from May through October; July and August account for 57% of the incidents.
- All events are between 12:00 and 9:15 p.m. MST.
- An equal number of males and females were affected.

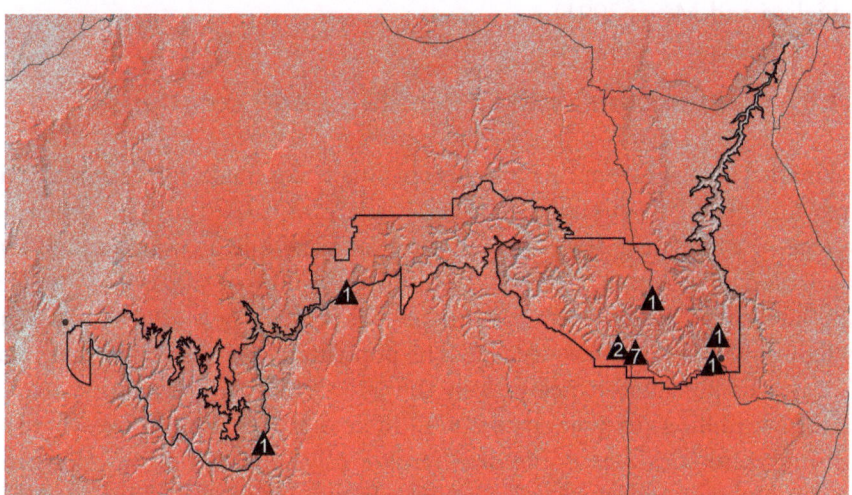

Fig. 5.3 Locations of 14 known lightning casualty events within Grand Canyon National Park overlain on point plot of 2009–2018 CG strokes and terrain. Triangles indicate the number of events at each location (updated from Holle et al. 2021). © American Meteorological Society. Used with permission

- Ages 20–29 and 40–49 were the most frequent decadal age groups.
- 77% of the reported origins of people were from outside Arizona.
- Events were spread widely throughout the week with no weekend preference.
- In terms of location when it was known, the South Rim was most frequent (9), below the rim in the Canyon was next (3), and one event each was on the North Rim and the Colorado River.
- In terms of activity, viewing the canyon from the rim was most common (6), while hiking (3), photographing (2), and guiding river rafters (1) were the other events with a known activity.

The *Arizona Daily Sun* newspaper in Flagstaff summarized an incident from more than 125 years ago that occurred in 1895 in its issue of August 9, 2020. This event took place before the park was established. A party of visitors from outside Arizona left a camp in the forest on an August morning to walk on a trail toward Bissell Point. At 11:00 a.m. when lightning occurred, they sought shelter under a large rock; one person was killed, and several others were seriously injured. While there certainly had been other deaths and injuries from lightning to Native Americans within what is now Grand Canyon National Park, this is the earliest event known to the authors. Similar more recent stories are included in the book *Over the Edge: Death in Grand Canyon*, and some of the details for the above summary come from this book (Ghiglieri and Myers 2001).

5.1.8   Comparisons with Other States and Locations

A remarkable result from this long-term casualty summary for Arizona is the large number of deaths in the earlier years from 1926 to 1929, when an average of more than seven people per year were killed by lightning. The population of the state in 1928 was only 422,000—compared with over 7,000,000 now. Yet less than two people per year are now killed in Arizona by lightning in the last few decades. The population-weighted rate, then, was 17.2 deaths per million per year in the late 1920s, then decreased to 0.2 deaths per million per year since 2000 (Table 5.3). We don't know what the lightning casualties were doing or where they were located during this early period, but some insight can be inferred from another article in Flagstaff's *Arizona Daily Sun* issue dated August 25, 2019. One hundred years earlier in 1919, one man was killed and two seriously injured by lightning; the three were Austrians chopping trees for a lumber company. They made the mistake of seeking shelter from lightning under two tall pine trees that were both said to be struck by lightning.

The greatly reduced rates in the last century are similar to those for the entire U.S. Fatality rates that have plummeted from over 6 per million per year in some years early in the twentieth century to under 0.2 deaths per million per year in recent years (Cooper and Holle 2018; Holle et al. 2005; Jensenius 2016; López and Holle 1996, 1998; Roeder 2012, 2016). The reduction is similar to the change in rural population involved in agriculture since then, as mentioned earlier. Among the most important changes affecting lightning exposure in the last century in the U.S. is that nearly all buildings where people live and work are now lightning-safe since they have grounded wiring and plumbing built according to building codes. In addition, some have lightning protection specifically installed in them for extra certainty to avoid interruptions to power and communications in critical operations such as airports, hospitals, police stations, and data centers (Sect. 5.3.1). Along with the greatly improved building practices is widespread access to fully enclosed metal-topped vehicles. As a result, most people in the U.S. most of the time are close, or very close to a lightning-safe building or vehicle. Other factors that contribute to this vast reduction in the lightning casualty rate are improved medical care (Cooper et al. 1999), the widespread availability of lightning information (Chap. 6), and persistent education about lightning safety that needs to be conducted over a long period of time (Holle and Zhang 2017; Jensenius and Franklin 2014; Roeder et al. 2012). The same reduction has occurred in such developed regions as Western Europe, Canada, Japan, and Australia where lightning-safe buildings and vehicles are nearly always nearby, and medical care, lightning information, and education are well established (Holle 2008a, 2016; Holle and Cooper 2019). There is also the unquantifiable fact that most people in the U.S. and these other countries spend now much of our time indoors where they are lightning-safe rather than enjoying the outdoors, such as in the remarkably varied terrain of Arizona.

Question

How much time do people spend outdoors these days of so many electronic devices? Is it more or less than 100 years ago? Does this factor contribute to the great reduction in lightning deaths and injuries in the U.S. and Arizona?

Note that the large fatality rate in the earlier part of the twentieth century in Arizona and the U.S. only applies in the developing world at the present time. Rates continue to be over 0.5 deaths per million people per year in partially developed countries but are in excess of 5.0 fatalities per million per year in much of Africa and presumably parts of Asia; data are severely lacking in most of these countries (Holle 2016). Here, the housing, schools, churches, and workplaces often have no grounded wiring or plumbing, there are no grounded metal structural components to carry a lightning strike into the ground safely, and properly installed lightning protection is rare. In addition, there are often few fully enclosed vehicles nearby for most people. The problem is especially acute since a large proportion of the population is involved in labor-intensive subsistence agriculture where there is no safe place to reach safety during mostly daytime work—when lightning most often occurs. Medical facilities are often poor, lightning detection data are rarely accessible, and lightning education is nearly non-existent (Holle and Zhang 2017; Raga et al. 2014; Roeder 2012). The book *Reducing Lightning Injuries Worldwide* by Cooper and Holle (2018) examines each of these factors in detail, specifically for the developing world.

Although we are quite safe from lightning in our everyday lives in the U.S. including Arizona, we can support organizations such as the African Centres for Lightning and Electromagnetics Network (https://aclenet.org/). Similar organizations are beginning to identify the magnitude of the problem in developing nations so that solutions can be found and potentially extended to South America and other regions (Cooper et al. 2016, 2018, 2019).

5.2   Lightning Safety in Arizona

We are pleased to include a new cartoon about lightning safety in Fig. 5.4 that was drawn specifically for this book by meteorologist and artist Wei Xu. The most important feature is showing that the person is running at the sight of lightning or hearing thunder. The second critical feature in the figure is to go inside a substantial building that is nearby. Also, Ms. Xu includes specific features of the Arizona landscape, namely the saguaro cactus and a large mountain in the distance. We discuss these ingredients in Chap. 3 in describing why Arizona lightning photos are so compelling.

5.2.1   Odds

What are the odds of a person being a victim of lightning? This question usually brings to mind a lottery ticket or some totally random and unlikely situation. You can't control whether you pick a lucky number in gambling.

Fig. 5.4   Arizona lightning safety cartoon (© Wei Xu)

But becoming a lightning victim is not random and not a matter of luck. There needs to be lightning at the same time and place as a person, and the person needs to be in a lightning-vulnerable situation. First, let's look at the situation for the entire U.S. as an example, using rounded numbers for simplification:

- 330,000,000 people in the U.S.
- 30 killed, 300 injured a year (Storm Data). 1 in 1,000,000
- Life span of 80 years. 1 in 12,500
- Major impact on 10 people. 1 in 1,250
- Summary: There is a 1 in 1,250 probability that a person in the U.S. will be killed or injured by lightning or be a close relative or friend of a victim, over an 80-year life span. The odds are somewhat better now, since the number of deaths is less than 30 and injuries are under 300 in recent years. The above numbers apply for a longer period than is now the case when lightning casualties continue to decrease in the U.S. (www.lightningsafety council.org). Nevertheless, it's apparent that the odds are not extremely large.

Taking this analogy to Arizona, we find the following, again using rounded numbers to provide a general rate:

- 6,830,000 people in Arizona.
- 1.4 killed, 5.3 injured a year from 2010 to 2019.
- That is a total of 6.7 casualties a year. 1 in 1,000,000
- Life span of 80 years. 1 in 12,500
- Major impact on 10 people. 1 in 1,250
- Summary: There is a 1 in 1,250 probability that a person in Arizona will be killed or injured by lightning or be a close relative or friend of a victim at today's casualty rate, over an 80-year life span.

It should be mentioned that if you, a family member, or an acquaintance is a casualty of lightning, there is the support group called LSESSI mentioned in Sect. 5.1.6. There you will find medical and emotional support for what to many can be a life-changing experience for the worse (Cooper and Holle 2018; Cooper et al. 2002).

So where does this discussion leave Arizona in the spectrum of possible lightning exposure to people? In the earlier parts of the twentieth century, the number of people killed and injured by lightning in Arizona was very large compared with the small population. At least some of the casualties were involved in manual labor-intensive agriculture, as well as mining and forestry. They were living and working in buildings that were not yet lightning safe, and there were few lightning-safe vehicles nearby. The knowledge of lightning was rudimentary, emergency medical treatment was sparse, and real-time lightning data did not exist. Since then, each of these factors has rapidly changed the situation until we now have few lightning deaths and injuries in Arizona that all take place outside a lightning-safe building or vehicle. We want to be outdoors, but during the afternoon in the monsoon-dominated months of July through September, vigilance is needed to be safe.

Occupational events had been frequent but the main contribution from manual agriculture is now mostly gone. In addition, the large reduction in wired landline telephones has diminished injury events inside the house. Outdoor recreation is now the largest category (Table 5.9). Vehicles are also a common source of lightning injury but nearly all of these events happen outside, rather than inside a fully enclosed metal-topped vehicle. Motorcycles and tents are especially dangerous. Going under trees when lightning is present in any scenario has never been, and still is never, a good idea.

5.2.2 Mechanisms of Injury

At this point, let's be sure to understand how lightning kills and injures. A more extensive version of the following summary in Cooper and Holle (2018) was targeted in part for developing nations and included a medical context, so here is a somewhat shorter version for Arizona.

As an electrical force, lightning can cause injury through five primary mechanisms. Their estimated frequency is shown in Fig. 5.5 and illustrations of each type are in Cooper et al. (2008) and Cooper and Holle (2018):

1. **Direct strike:** There is nothing between the person and the lightning that contacts (or 'attaches to') them. In developed countries, it is estimated that only 3–5% of fatalities are caused by a direct strike.
2. **Contact voltage:** Lightning hits something else first and travels through that pathway to affect someone who is holding onto the energy transmitter. Examples are turning on a water faucet when lightning has hit the ground a distance away and been transmitted through the water or plumbing; or someone talking on a hard-wired phone (Andrews 1992; Andrews and Darveniza 1989). It is estimated that contact injury causes approximately 15–20% of deaths in developed countries.
3. **Side flash or splash:** Lightning hits another object and a portion of the energy jumps to a nearby person to complete its path to ground. An example is someone standing under a tree. About 20–30% of lightning deaths are caused by a side flash.
4. **Ground current:** This may also be called step voltage, ground potential, and other terms. In this situation, lightning hits the ground a distance away from a person and spreads through the ground nearly radially. It

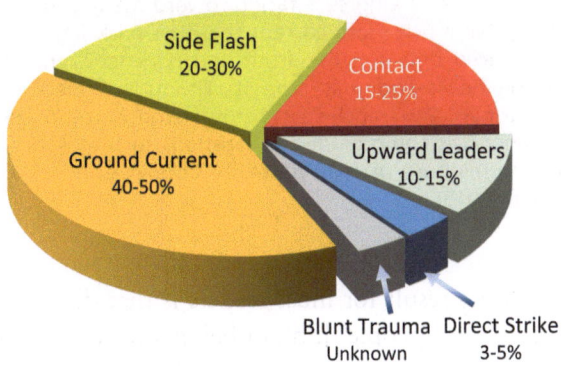

Fig. 5.5 Mechanisms of lightning injury

may go up one leg and down the other in a standing person or from head to foot in someone sleeping in a tent. It is thought to cause 40–50% of deaths in developed countries, but further examination shows that ground current may not be this common (Andrews 2021). Ground current can be divided into two groups:

a. *Step Voltage* occurs when the energy stays in the ground as it contacts the person and returns to the ground from the person.
b. *Ground Arcing* occurs when the energy jumps through the air such as across the mouth of a cave where a person is sheltering.

5. **Upward leader**: The term upward streamer also applies to this mechanism. Any electric field, such as in a thunderstorm, induces an opposite charge in objects under it, including tall towers, trees, people, or blades of grass. These charges can form an upward leader from the object toward the strong overhead electric field. A vivid example is shown over the desert of southwestern New Mexico by Cummins et al. (2018). The streamer may not attach to the downward leader of lightning from the overhead thunderstorm, but it has enough energy to injure a person (Anderson 2001; Anderson and Carte 2009; Anderson et al. 2002; Cooper 2002). Upward streamers are thought to cause about 10–15% of lightning deaths, although Andrews (2021) raises the question of whether they may be more frequent.

Large Animals

One of the worrisome factors related to ground current is the danger to four-legged animals. When lightning's current spreads outward across the ground, it weakens with distance. If a horse is facing toward the ground strike point, its front legs will receive much more current than the back legs. A large voltage difference is produced between the front and back legs, and the difference goes through the heart and may stop it. This is not merely an academic exercise, since farmers and ranchers can lose large herds of cattle in these situations. The impact is compounded in locations such as Africa where cattle are sometimes the measure of family wealth (Cooper et al. 2017; Cooper and Holle, 2018; Elsom 2015).

The most surprising result for most readers is that direct strike is actually rare, yet this is how most people think of being affected by lightning. Forget the old clichés about holding a golf club or fishing rod. Direct strike hardly ever happens, but instead, many other mechanisms are more dangerous. As a result, lightning safety needs to be considered more completely than simply lowering oneself. There's only one answer in Arizona—go inside a lightning-safe building or vehicle. Ignore the rest of the false advice of how you stand, what you hold, or what you are wearing (Cooper and Holle 2011). As shown on the front page of the National Lightning Safety Council website at http://lightningsafetycouncil.org/: "NO PLACE OUTSIDE IS SAFE when thunderstorms are in the area". And most importantly, plan ahead so that you know where to reach a safe place. When lightning is nearby, it is too late to ask or think calmly about where to be safe.

The Crouch Position Is Irrelevant

A sign that a set of lightning safety advice is mistakenly mentioned is the lightning crouch. In this idea, lowering oneself to reduce the lightning impact is the first recommendation in older policies. But, since the direct strike has been found from examining many reports to be the least likely mechanism of lightning injury, it should be the last mechanism to consider. The mistaken idea that lightning comes down from above to strike a person directly on the head is a popular but mistaken idea. Roeder (2014) calculated that lowering oneself by half a person's height reduces direct strike exposure by perhaps a half. So, it reduces the 3–5% direct strike to 1.5–2.5%. Is that worth the emphasis? No. Instead, in the presence of lightning, go as fast as possible to a lightning-safe building or vehicle.

People can also be injured by several non-electrical mechanisms in the case of lightning, as in the following three methods (Cooper and Holle 2018):

1. **Shrapnel or missile injury**: A person receives penetrating trauma as lightning blasts shrapnel such as tree bark, or material is thrown from a cement walkway into them (Blumenthal 2012).
2. **Blunt injury**: A person is thrown a distance by muscular contraction induced by lightning. This situation can result in trauma similar to a fall, often with musculoskeletal injuries.
3. **Barotrauma**: This can be in the form of a concussion, blast, blunt force, or explosive injury. When a person is close enough to the lightning channel to experience a rapid outward movement of air, the situation is similar to an explosion that may be strong enough to be knocked off their feet or cause concussive injuries to internal organs. An investigation by a forensic pathologist calculated the pressure wave from lightning at a distance of 10 m from the strike to be similar to the force of a 5 kg TNT bomb (Blumenthal and West 2015).

What do these mechanisms have in common? They are nearly all affecting people outside the protection of a well-constructed building or a fully enclosed metal-topped vehicle. The most dramatic situation is at the Grand Canyon where millions of people visit and tour outdoors, and lightning can kill or injure someone. Similarly, hiking and climbing on Arizona's magnificent mountains can be very dangerous.

A common misconception is that people inside a car are safe because it has rubber tires. When lightning strikes a car, truck, bus, or other fully enclosed metal-topped vehicles, the current travels around the outside of the vehicle, making people safe inside. This myth apparently resulted from the fact that sometimes tires are flattened when a vehicle is struck. The typical path is when lightning hits the roof or another part of the car, and travels through metal since it is good at conducting electrical current. It then reaches the lower part of the car, including the axles, which have wheels attached, so a fast way to ground is through the tires which may not withstand the massive surge of current from lightning! The tires did not save a person inside, it was the metal Faraday cage surrounding the individual. The blown tires, then, are an effect of the lightning strike not a cause of protection.

Human Impacts

Lightning injuries are primarily neurological, namely, spinal and brain injuries, making them difficult to diagnose and treat. The effects have some similarities to Post-traumatic stress disorder (PTSD).

5.2.3   Safe and Unsafe Places

Not all lightning in Arizona occurs in the presence of rain, however, the typical reaction is to pay more attention to the rain rather than the lightning. Rain causes discomfort and makes people and objects wet but does not normally injure unless flooding occurs. In contrast, lightning is very intermittent in time and space but is potentially fatal.

There are only two reliable places to be safe from lightning. These are (1) inside a large substantially constructed building and (2) inside a fully enclosed metal-topped vehicle. These locations provide an effect similar to a Faraday cage such that lightning striking a building or vehicle travels around, rather than through the people inside it. Let's examine these in more detail:

(1) *Buildings*: Lightning-safe buildings have paths for the flash to follow through grounded wiring and plumbing in the walls when they are properly installed according to accepted municipal building codes. They may have metal structural members that are part of the building itself such as metal studs (Holle 2010). Direct strikes to dwellings with people inside are quite common in locations such as Arizona, but rarely result in fatalities. Injuries are mostly minor, although they can be very undesirable, due to people being in contact with the conducting paths of wiring and plumbing. Note that the metal is not attracting lightning but provides a safe path for the current surge from lightning to follow. A special case to mention is that some buildings have lightning protection installed such as lightning rods. This lightning protection is located at airports, fire and police stations, hospitals, manufacturing plants, and other facilities with critical infrastructure, just to be certain that no lightning current enters the building to interrupt sensitive operations (Sect. 5.3.1).

Assume any other building is unsafe—especially small structures. It is possible to make them lightning-safe, but it takes a specialized effort by a licensed experienced specialist in lightning protection and incurs expenses that may not be cost-effective or practical (Gomes and Izadi 2019). As a result, assume that a small and/or open-sided structure of any type is not safe. These include agricultural outbuildings, shacks and huts, roadside shops, sun shelters, beach shelters, rain shelters, golf shelters, bus shelters, forest huts, snack shops at small parks and recreation centers, and similar small enclosures (Sect. 5.3.1).

The 1797 San Xavier Mission near Tucson is almost continually under restoration, currently on its East Tower. Lightning safety advice is shown prominently at the entrance (Fig. 5.6). The location of workers at a higher elevation than the surrounding area makes them susceptible to lightning. However, it is not the metal itself but the isolation higher than the nearby flat land. We are not aware of the method used for warning.

Church Bell-Ringer Deaths from Lightning

This San Xavier Mission lightning safety advice is ironic due to lightning problems at churches before Benjamin Franklin invented the lightning rod. Church bell ringers in Europe were sent to towers to ring bells, thinking that their sound could ward off lightning during thunderstorms. Unfortunately, it was reported that during a 33-year period, 103 people were killed and 386 church towers were struck. This number is quoted by Schonland (1950, page 8) from a 1784 Munich, Germany book by Fischer with the title "A proof that the ringing

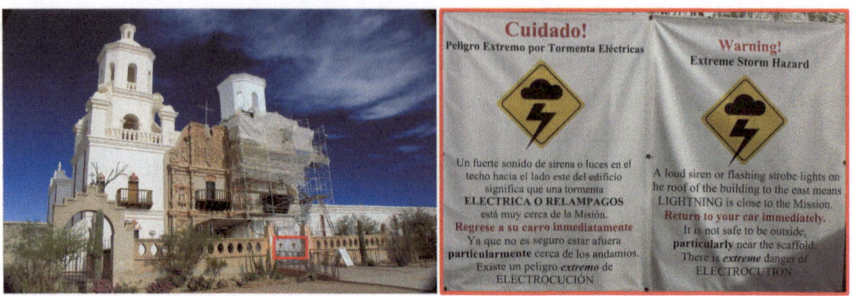

Fig. 5.6 Lightning safety instructions at San Xavier Mission south of Tucson during the renovation of the East Tower (© R. Holle)

of bells during thunderstorms may be more dangerous than useful." This study is mentioned in less detail by Tomlinson (1848, page 215), and De Fonvielle (1869, page 169) found it necessary to include a chapter titled "Storms are not affected by bells." These numbers are also quoted more recently by Uman (1986).

(2) *Vehicles*: Lightning-safe vehicles are fully enclosed and have metal tops. A direct strike can be very disconcerting, but people always survive a lightning strike while inside them (Holle 2008b). In practice, they can be considered as portable lightning safety locations that can be moved to critical locations when the threat of a thunderstorm arrives. Lightning-unsafe vehicles include golf carts, cloth-topped, four-wheelers, and similar open-sided vehicles, in other words, any vehicle that does not surround people who are inside with a metal cage-like effect. The effective and simple lightning warning sign at Utah's Bryce Canyon National Park in Fig. 5.7 is applicable to the Grand Canyon and other parks, mountain slopes, and rims throughout Arizona.

Two other situations deserve special attention that are often considered with respect to lightning, trees, and tents:

(1) *Trees*: Around 10% of all lightning casualties occur under trees. The immediate reaction is to try to stay dry under a tree, but a very real lightning threat occurs here (Mäkelä et al. 2003; Holle 2012). The percentage of people killed and injured under trees has been in the 10% range for decades, in all countries. Animals also tend to huddle under trees from the rain and thereby have a significant risk from lightning, as observed

Fig. 5.7 Warning sign for lightning at Bryce Canyon National Park (© R. Grad)

from the large number of reports of domestic and wild animals found dead under trees (Cooper and Holle 2018).

(2) *Tents*: They provide rain protection but are not lightning-safe. The problem is compounded when thunderstorms occur at night (Holle 2014). Similar situations include going under open-sided stands at soccer or other recreation venues to escape rain while the lightning threat prevails (Holle 2005). A large tent was the location of a tragic lightning event involving girl scouts in South Africa (Cooper et al. 2015).

5.2.4 Actions Before and After Thunderstorms

"When thunder Roars, Go Indoors" is a phrase that was developed by Mr. William Roeder and is a one-step way to manage the lightning risk (Roeder 2009). This phrase has been paraphrased to rhyme with the same message in Spanish and is known globally as part of lightning education (Yas et al. 2021). In particular, the National Lightning Safety Council has it as one of its core messages displayed on its homepage at http://lightningsafetycouncil. org/. This message provides a clear indication that if thunder is heard, go inside a lightning-safe building or vehicle. As mentioned in Chap. 1, thunder has a range of distances where it can be heard. In general, thunder can be heard within a few miles of where lightning is occurring. However, man-made noise such as traffic, as well as rain, and wind blowing through trees may make it nearly impossible to hear thunder when you're outside. Since sound travels about 1 mile in 5 s (Chap. 1), it takes 30 s for sound to reach you from lightning 6 miles away, and that is close enough to take immediate action. It goes without saying that light travels so fast that you have no time to react to the light from a flash in terms of safety, but we can be thankful that thunder provides some warning!

How long do you wait until it is safe to return outside? A wait time of 30 min is usually adequate. The result then is the 30–30 rule first developed at a 1998 conference (Holle et al. 1999). Until this rule was established, there had not been an objective way to provide safety advice for both the start and end of the lightning threat. The following extract from Holle et al. (1999) summarizes the outcome:

> A major result of this meeting was a general agreement concerning the '30–30 rule.' The first 30 refers to the number of seconds between 'flash' and 'bang' that initiates safety precautions. The second 30 refers to the number of minutes after the last flash or thunder in order to establish an all-clear signal.

Practical implementation of the 30–30 rule is clearly problematic in some situations, but it is a starting point. For example, in an active storm, which CG stroke or in-cloud (IC) pulse is associated with which sound of thunder? If you use "When Thunder Roars, Go Indoors", you don't need to pay any attention to details, just go inside and don't ask questions. However, when more nuanced lightning avoidance is needed, consider an airport. It has been calculated that delay costs are very large when a lightning alert is issued, then outdoor workers stop activity and passengers are not able to leave or enter an aircraft. Finding the correct balance between lightning safety risk and traffic ground delays is difficult, since the cascading costs for a ground delay can become huge (Steiner et al., 2013, 2016). As a result, variations of the 30–30 rule are used around the world for aviation, mining, recreation, and other vulnerable industries. Consider the start and end of the lightning threat from a thunderstorm:

- **Start:** At the start of the storm, 10 km (about 6 miles) is a commonly used range for when lightning triggers work stoppage. If the area being monitored is much smaller than 10 km, lightning can suddenly appear on top of you without enough time to take action.
- **End:** At the end of a storm, the 30–30 rule says to resume outdoor activities after 30 min. In practice, most operations wait 10 or 15 min, but making the wait time too short runs a risk of having to stop and start too often. The larger the operation and the longer the evacuation time, the longer the wait time at the end of a storm needs to be. For example, if you are in the backyard planting a cactus and hear thunder, go inside, and wait 10 min after the last lightning or thunder, since you can go back

inside by taking only a few steps that takes seconds of time. But if a large outdoor event is taking place so that it takes half an hour to move people to lightning-safe buildings or vehicles in a parking lot, wait 30 min after the last lightning and thunder in the vicinity. There is nothing worse for building trust in credible warnings for lightning safety than to have thousands of people streaming back and forth over short periods! Although the 30–30 rule is widely used, extremely large and long flashes occur as mentioned in Sect. 1.4.5, so there may need to be a stricter rule for areas with large Mesoscale Convective Systems (MCSs) such as sometimes occur in Arizona at night during the monsoon.

All of these parameters are adjustable and are explored in extensive studies such as Holle et al. (2016). Remember that the presence of nearby IC lightning is equally as important as CGs. Most of the time, there are several times as many ICs as CGs, and they serve as a critical indication that lightning may come to ground at any time. All lightning warning systems such as considered in Holle et al. (2016) include ICs with CGs and their value is readily evident.

30–30 Rule for Children

The 30–30 rule became so popular when it was introduced that a children's book was written using animals, artwork, and a brief text to explain it (Florida Division of Emergency Management 2004). The seconds are counted by the name of the state capital, as One Tallahassee, two Tallahassee', etc. The first author attended Florida State University there and finds this to be quite clever.

One situation has not yet occurred in Arizona but may in the future—that is during high school football on Friday nights. Dozens of games are played in the early evening from late August through September into October when lightning can occur. An afternoon UArizona football game had a significant and difficult delay as reported by Holle and Krider (2006). Lightning safety rules need to be developed for each stadium and followed to avoid a major incident (Walsh et al. 2013).

One further thought on lightning safety. Paul Holle, younger son of the first author, is an instructor for the National Outdoor Leadership School and teaches lightning safety as part of his course. One of the underlying principles of coping with threats in the outdoors is that medical emergencies are due to a series of seemingly inconsequential decisions that culminate in being at the wrong place at the wrong time. A book describing this process is Gonzales (2004).

Question

What are some of the seemingly small decisions that can lead to lightning casualties while climbing a mountain, playing a football game, working on a utility pole, or standing under a tree?

5.3   Lightning Damages in Arizona

Lightning damages are notoriously elusive to quantify. There are two categories of expenses, short-time and long-term. In the shorter time period, power is shut down, the internet and cable go down, houses and trees are damaged or destroyed, and forest fires are started. In the longer time period, avoidance is the goal so that power and communication lines are protected, and they do not cause so many of these interruptions.

It should be mentioned that lightning death and injury data in Sect. 5.1 are complete through 2021, since such events are widely known. However, damages are reported less often than casualties, so we rely on *Storm Data*.

5.3.1   Lightning Protection

Where have you seen lightning rods in Arizona? What were they protecting? Did you notice that most large power poles have a wire over the top of the transmission or distribution lines? Those are called shield wires installed specifically so that lightning does not reach the main electric lines below that are conducting electricity from one location to another. As you drive down the street or road, notice how often the shield wires are on top of the utility poles next to the road but you didn't notice them before...but first pay attention to your driving!

The long-term costs of installing protection are probably greater than the immediate losses. Power lines, communication towers, emergency power at hospitals, airports, police and fire stations, along with traffic lights, and home Wi-Fi networks are all vulnerable to lightning, but we don't want such interruptions, ever. Protection costs money, lots of it, very likely in the hundreds of millions of dollars or more in Arizona every year. But there is no way to know these costs because they are built into the design of every part of our infrastructure. Proper lightning protection is a sophisticated engineering discipline with constant demands on eliminating any disruption, and there are ever-expanding applications to manage new lightning risks to such facilities as wind turbines and solar panels. Very large global conferences such as the International Conference on Lightning Protection (http://www.iclp-cen tre.org/) are held frequently, and comprehensive books have been written on the topic such as Rakov and Uman (2003).

What if you are a homeowner or apartment dweller in Arizona? You most likely do not have lightning protection specifically on your dwelling. However, it is provided by grounded wiring and plumbing for your building that was installed according to standards established for construction in your town, city, or county. A sure sign that something is not correct is if some part of your home often has outages due to lightning. Although the whole house is protected by building code standards, you can take two additional steps. One, install surge protectors at each vulnerable location such as a computer or other major electronic device. These need to be more than a terminal strip but designed to withstand a direct impact on the home or apartment. Much, or most lightning damage to dwellings comes through the incoming power so that protecting against a surge is the most straightforward way to proceed. Two, you may be especially vulnerable to lightning if you have a computer-based home office or are in a rural area with few neighbors nearby. If so, install lightning protection on the house itself directly onto the incoming power line. Use a bonded insured experienced electrical installer using nationally accepted standards.

So why were there lightning rods on barns in the Midwest where many Arizonans originated? The barns did not have enough wiring and plumbing to provide lightning protection—there may have only been a single power line coming to a few lights or small machines inside the barn, but that's not enough. If you look carefully at present-day buildings such as airport terminals, fire stations, and hospitals where power should never go out, you will see small rods that are not as fancy as they used to be, but still do the job. There are three essential parts to a successful lightning protection system on any structure:

- *Rod*: Isolated substantial rods that are taller than the surrounding. They need to be made of metal so that they do not disintegrate when struck and will carry the massive short-term surge from lightning into the ground. They do not need to be especially tall, but if they are short, a larger number of them is needed.

- *Down conductor*: Carries the surge from the rods down toward the ground. All lightning rods need to be connected by very thick metal cables.
- *Ground rod*: A very sturdy thick pole is pounded into the ground so that it dissipates the lightning current into the dense earth. There may be several ground rods, or a trench with grounding conductors may be installed in some situations to connect with the down conductor.

Mirrors

In many cultures, covering mirrors is considered to be a way to prevent lightning (Elsom, 2015). While it may sound strange, we have heard this myth in parts of North America, Asia, Europe, and Africa. For example, in China, a myth says that the lightning mother holds a mirror in each hand, and shoots lightning before the thunder god shoots thunder, so that he can see clearly (Zhang 2022). It's difficult to understand any logical reason for this thought process! What do you think could be a reason?

5.3.2   Storm Data

Storm Data was used extensively in Sect. 5.1 to describe lightning deaths and injuries. This publication also has damage reports; however, it is not nearly as complete regarding other information as for human casualties. Lightning reports from 1959 through 2019 in *Storm Data* tend to occur in these categories:

1. Tree damage, especially palms.
2. Houses set on fire.
3. Buildings.
4. Utility damage resulting in power outages and failed equipment such as utility poles.

Damage reports due to lightning in *Storm Data* are subject to a major limitation. There is no way for National Weather Service staff to know all of the outages, internal home damages, and interruptions that occur. Each insurance company keeps its own data, and lightning claims are often paid directly with minimal or no reporting to anyone else. In fact, a study in Colorado, Utah, and Wyoming from 1987 to 1993 found that 367 times as many lightning

Table 5.16 Decadal totals of the number of lightning-related damage reports in Arizona from 1959 through 2019. Most reports are from *Storm Data*, and a few are from newspaper articles and web resources

Period	Events
1959–1969	14
1970–1979	27
1980–1989	37
1990–1999	32
2000–2009	64
2010–2019	77
Sum	*251*

claims were actually paid to insurance customers in this study period as were included in *Storm Data* (Holle et al. 1996).

5.3.3 Decadal Changes in Damage Reports

By decade, Table 5.16 shows a general increase in damage reports since 1959. The number of events is quite small in the early years, and steadily grows. Most of the reports are from *Storm Data*; however, there is a very noticeable variability in reporting. It seems that for a while either the Phoenix or Tucson National Weather Office became focused on such impacts as utility outages, and in other years on trees. Both types of events occur often, so these reports are a result of a policy change or emphasis that took place for a few years in most cases. They only represent a small portion of the actual events. As a result, the data shown here may be representative of the types of damage but are not comprehensive (Holle et al. 1996). As a result, it is not possible to extrapolate data in Table 5.16 to any actual rate of lightning-related damages; such data are closely held by utility and insurance companies.

5.3.4 Monthly and Hourly Variations in Damage Reports

Table 5.17 indicates a strong concentration of damage reports in July and August, comprising over two thirds of the year's total. September follows in frequency, and the rest of the months contribute minimally to the annual total of lightning damages.

By time of day, Table 5.18 shows information on when damages are reported from various sources. A number of reports indicate only the general time such as afternoon, while others are exact. When "afternoon" events are

Table 5.17 Month of 251 lightning-related damage events, and percent of annual total across Arizona from 1959 to 2019

Month	Events	Percent of annual total
January	4	1.6
February	4	1.6
March	5	2.0
April	2	1.6
May	10	4.0
June	9	3.6
July	83	33.1
August	97	38.6
September	26	10.4
October	10	4.0
November	1	0.0
December	0	0.0

Table 5.18 Time of day in Mountain Standard Time of 208 lightning-damage events, and percent of total within Arizona from 1959 to 2019

Time	Events	Percent
General		
Afternoon	6	2.8
Evening	5	2.4
Specific		
Midnight to 0600 (midnight–6 a.m.)	20	9.6
0600–1200 (6 a.m.–noon)	7	3.4
1200–1500 (noon–3 p.m.)	22	10.6
1500–1800 (3–6 p.m.)	56	26.9
1800–2100 (6–9 p.m.)	56	26.9
2100–Midnight (9 p.m.–midnight)	36	17.3

combined with those from 12:00 to 6:00 p.m., 40% of the events are found to be during this period. The outlier is the 1200–1500 time period with frequent lightning but fewer damages than expected. This is much less of an afternoon concentration than the 60% found for deaths and injuries. It can be conjectured that people are more likely to be exposed to lightning while outside during daylight hours, while the lightning itself lingers into the evening and nighttime hours to result in a rather large number of damage reports after 6:00 pm (18:00).

Correlations of CG flashes with months and time of day of casualty events are shown in Fig. 5.8. The correlation is very high on the monthly scale ($r = 0.9748$). However, there is a much weaker correlation for time of day ($r = 0.3851$). The time of day also had a very good correlation for deaths and injuries (Fig. 5.2); however, for damages, it is much less clear.

Question

What can account for the much weaker correlation of damages with time of day than human casualties? Objects don't move, so how can there be so many lightning death and injury events in the afternoon in Table 5.5 and Fig. 5.2 but so few afternoon damages to objects such as trees, houses, buildings, and utilities? Is it possible that damages are not noticed during the afternoon while storms are occurring? It's not at all clear what is happening here! Can you suggest other reasons?

5.3.5 Damage Reports by County

Where are these damage reports? County information from mostly *Storm Data* in Table 5.19 shows Maricopa County to lead the way, followed by Pima and Mohave. Maricopa and Pima were by far the most populous counties during the data period since 1959.

Ranking

The ranking of Mohave County as third is not an obvious choice, although it has less than half as many damage reports as Pima. Reports reached *Storm Data* more often from Mohave than most other counties for unknown reasons— Mohave County is fifth in population behind Pinal and Yavapai at present. Another confounding factor is that the amount of lightning is quite small in Mohave County (Fig. 4.3).

5.3.6 Types of Damages

A distinctive reported damage is to trees (Table 5.20); all of these are in urban areas. Palm trees are often reported, sometimes in groups, but it is not known if they were simultaneous or in a neighborhood during a thunderstorm. These

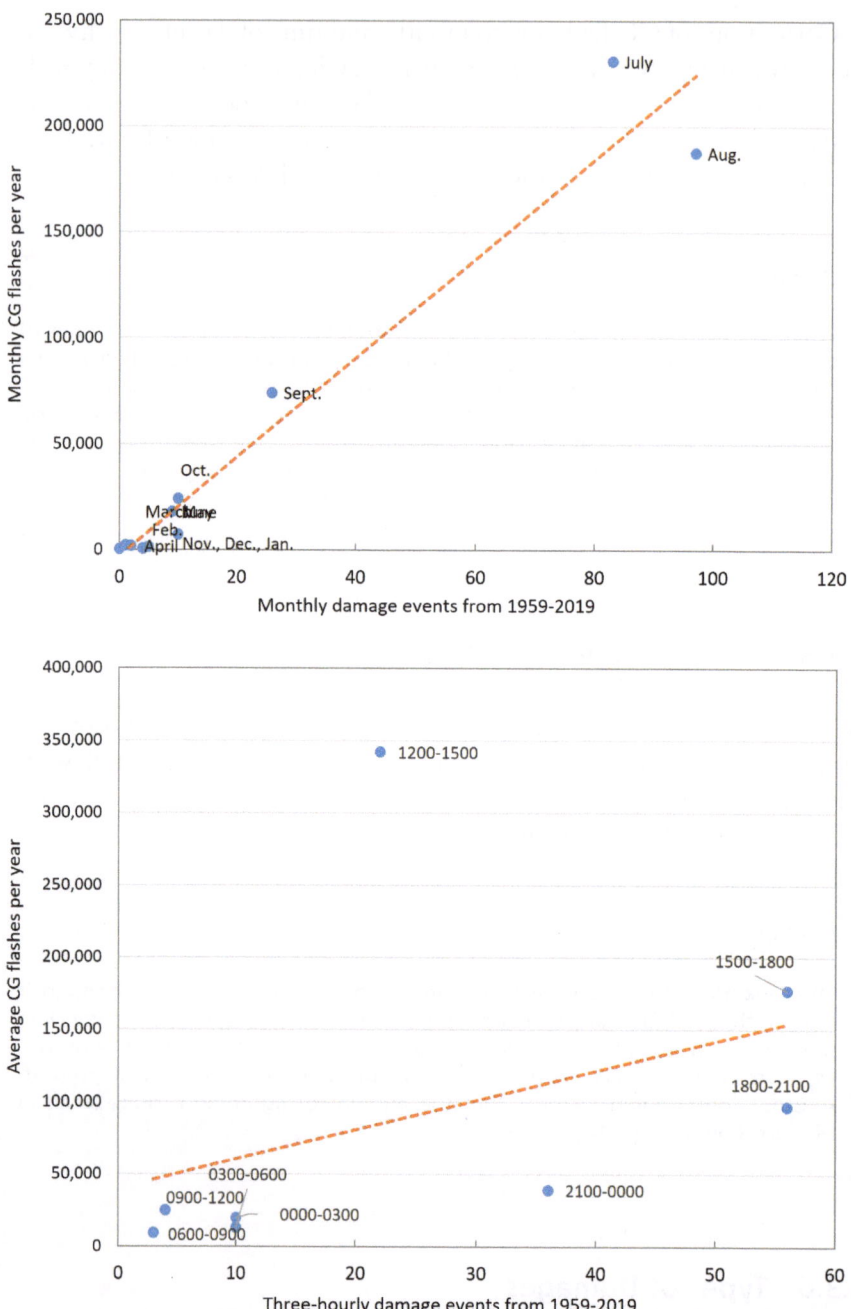

Fig. 5.8 Top panel: Correlation of monthly damage events with monthly CG flashes per year. Bottom panel: Correlation of 3-hourly damage events with 3-hourly CG flashes per year

Table 5.19 Counties where 251 lightning-related damage events occurred, percent of total, and rank within Arizona from 1959 to 2019

County	Number	Percent	Rank
Maricopa	101	40.2	1
Pima	66	26.3	2
Mohave	28	11.2	3
Yavapai	19	7.6	4
Gila	9	3.6	5
Cochise	8	3.2	6
Coconino	5	2.0	7
Pinal	5	2.0	7
Yuma	3	1.2	9
La Paz [since 1983]	2	0.8	10
Navajo	2	0.8	10
Apache	1	0.4	12
Graham	1	0.4	12
Santa Cruz	1	0.4	12
Greeenlee	0	0	15

Table 5.20 Lightning-related damage events related to trees within Arizona from 1959 to 2019

Type of tree	Events
Palm tree(s)	
One palm	6
Two palms	3
Three palms	1
Three palms on fire	3
Four palms	1
Eight palms	1
Several palms	1
Many palms on fire	1
Palms struck, blew out windows at duplex	1
Other trees	
One tree	5
Tree fire	2
Eucalyptus exploded	1
Numerous tree fires	1
Several tree fires	1
Tree split	1
20 trees on fire	1

cases are mostly from Maricopa and Pima counties at elevations low enough to have palms in their urban areas.

Speaking of trees, what about saguaros? They make beautiful silhouettes for Arizona lightning photos (Chap. 3) but are they struck by lightning? Indeed, they are, since they are often the tallest and most isolated plant in the lower desert that normally has only small trees or no trees at all. However, the vast majority of saguaros are on public lands, not in yards, so there are no reports of saguaro damage in *Storm Data*. Steenbergh (1972) found that lightning in a 1969 thunderstorm in the Tucson Mountains west of Tucson killed 11 plants in a small area. Later, Noggle et al. (2002) examined a 2000 lightning strike on a saguaro to find that the current flowed outward from a struck plant up to 5 m (tens of feet) over a very rocky surface.

An unexpected category has seven *Storm Data* events involving hay (Table 5.21). Much of rural Arizona continues to be oriented toward agriculture, so these events occurred across the state throughout the time period of record. The cost can be quite large; dollar amounts are not adjusted for inflation.

Turning now to damages to the built environment, events are divided into homes (Table 5.22), buildings (Table 5.23), transformers (Table 5.24), power outages (Table 5.25), and other infrastructure (Table 5.26).

Homes account for a large number of lightning damage reports (Table 5.22). Around half of them involve fires, and quite a few are destroyed or badly damaged. Starting in 1996, 75% of lightning-caused home damages were due to fires. An unusual feature is that multiple homes are sometimes involved in one report from a single day in a specific region. These may be due to incoming power surges from lightning hitting a power pole or other wired infrastructure spreading to several homes, however, this can only be determined by direct inspection at the site by qualified fire department and/or insurance investigators. Note that seeking safety inside a home from lightning is an excellent solution when lightning is in the area; homes are often affected by lightning, but people can escape (see lightning safety in Sect. 5.2). The very rare fatalities inside U.S. homes in recent years have always occurred at night when an elderly and/or mentally or physically compromised person is unable to escape the fire.

Table 5.21 Damage events related to hay within Arizona from 1959 to 2019. Values are not adjusted for inflation

Category	Year	Value
360 tons	1963	$ 27,500
Hay	1975	$ 21,000
Row ¾ mile long	1978	?
400 tons	1984	?
500 tons	2006	$100,000
Haystack	2009	$ 10,000
Large stacked bales	2016	$100,000

Table 5.22 Homes impacted by lightning within Arizona from 1959 to 2019

Category	Events
One home per event	77
House fire	39
Destroyed	16
Home/house	16
Moderate damage	3
Extensive damage	2
Roof damage	1
Multiple homes per event	20
Two houses on fire	6
Several	5
Two houses	3
Four houses	2
Six houses	2
13 houses	1
Three houses	1
Mobile/manufactured homes	5
One home	2
Fire	2
Destroyed	1
Appliances in homes	3

The other types of buildings damaged by lightning are diverse (Table 5.23). Stores (9 events), apartments (8), and others (8) are the most common. The fact that these events are widely varied shows that the vulnerability to lightning is very broad.

Electric power impacts are specifically called out in Tables 5.24, 5.25, and 5.26 since they affect incoming lines that can affect structures of all types. In addition, power reliability is expected in these days of continuous connectivity at home and elsewhere, such that power outages of any type are not tolerated. Notice in Table 5.23 that a nuclear reactor was shut down due to lightning in one event. There is no question that the transformer events in Table 5.24 are an undercount since most reports do not reach the National Weather Service to be entered into *Storm Data*. The outages in Table 5.25 are also undercounted since they are not unusual in the metropolitan areas of Phoenix and Tucson but are not entered into *Storm Data*. Other events in Table 5.25 apparently affected so many people that they were included in *Storm Data*.

Table 5.23 Buildings impacted by lightning within Arizona from 1959 to 2019

Category	Events
Apartments	8
Apartment complex	2
One apartment damaged	2
Two attached apartments damaged	1
One apartment ·	1
One apartment destroyed	1
One condominium	1
Manufacturing	5
Plastics manufacturing	2
Factory	1
Industrial building	1
Welding shop	1
Medical	2
Hospital	1
Medical office	1
Police/military	4
Police station	3
Air Force Base fire department	1
Radio/TV stations	4
Radio station	2
Antenna	1
Three TV stations off air for hours	1
Schools	2
Elementary/considerable damage	1
High school	1
Stores	9
Drug store	2
Gas station	2
Large store	1
Laundry	1
Market	1
Restaurant	1
Store	1
Other	8
Building	3
Shed	2
Brush fire burned down building	1
Nuclear power plant; one reactor shut down	1
Resort	1

Table 5.24 Damage events related to transformer strikes by lightning within Arizona from 1959 to 2019

Category	Events
Power line/transformer	8
Tranformer fire	2
Numerous transformers	1
Transformer caused hospital evacuation	1
Transformer led to house fire	1

Table 5.25 Power outages due to lightning within Arizona from 1959 to 2019

Category	Events
Power outages	17
>0 to 1,000 customers	5
>1,000 to 5,000 customers	4
>5,000 to 10,000 customers	1
>10,000 to 25,000 customers	3
Unknown	4
Power lines down	4
Power poles struck	2
Large area of outage	1
Outage lasting three hours	1
Power pole struck, started roof fire	1

Table 5.26 Other infrastructure lightning damages from within Arizona from 1959 to 2019

Category	Events
Considerable damage	1
48 kW circuit breaker at generating plant	1
Nuclear power plant struck, one reactor shut down	1
Pager tower	1
Private communications system	1
Ruptured gas line	1

5.3.7 Costs of Damages

Storm Data damage costs are evaluated and summarized based on two very different systems, such that estimates need to be taken into account separately to determine actual costs in current U.S. dollars:

- 1959–1995: *Storm Data* used the category system for damages in Table 5.27. For lack of a better approach, the average of the largest and lowest

Table 5.27 Damage cost categories for *Storm Data* events from 1959 to 1995

Category	Range	Average
1	<$50	$25
2	$50 to $500	$275
3	$500 to $5,000	$2,750
4	$5,000 to $50,000	$27,500
5	$50,000 to $500,000	$275,000
6	$500,000 to $5,000,000	$2,750,000
7	$5,000,000 to $50,000,000	$27,500,000
8	$50,000,000 to $500,000,000	$275,000,000
9	$500,000,000 to $5,000,000,000	$2,750,000,000

costs in each category is used as shown in the right column of the table, although this is not especially satisfactory. Inflation adjustments are made to December 2020 by using the Consumer Price Index (CPI) Inflation Calculator at www.bls.gov/data/inflation_calculator.htm.

- Starting in 1996: Actual amounts were listed in *Storm Data*. Since there also are inflation factors to take into account for this period, inflation adjustments are made to December 2020 with the CPI Inflation Calculator at www.bls.gov/data/inflation_calculator.htm.

The known costs for events in Tables 5.20 through 5.26 are combined in Table 5.28 into two categories of homes and everything else:

Table 5.28 Lightning damage costs for homes and other buildings and objects from *Storm Data* from 1959 to 2019. Inflation-adjusted using CPI Inflation Calculator at www.bls.gov/data/inflation_calculator.htm

Damage range	Homes	Buildings and objects
$0 to $1,000	3	22
>$1,000 to $5,000	3	11
>$5,000 to $10,000	10	8
>$10,000 to $25,000	9	14
>$25,000 to $50,000	7	6
>$50,000 to $75,000	8	5
>$75,000 to $100,000	8	0
>$100,000 to $250,000	24	7
>$250,000 to $500,000	2	0
>$500,000 to $750,000	3	1
>$750,000 to $1,000,000	1	2
>$1,000,000 to $2,500,000	0	2
>$2,500,000 to $5,000,000	0	1
>$5,000,000 to $125,000,000	0	1

Table 5.29 Ten largest lightning damage incidents from *Storm Data* from 1959 to 2019. Inflation-adjusted using CPI Inflation Calculator at www.bls.gov/data/inflation_calculator.htm

Damage range	Damage
$116,000,000	Trees in forest fire
$4,827,000	Manufacturing plant
$2,364,000	Homes and factory
$2,333,000	Drug store
$1,302,000	Agricultural machinery
$942,000	Home
$868,000	Apartment building
$666,000	Home
$651,000	Home
$632,000	Plastics manufacturing company

- **Homes** (Table 5.28): They have an inflation-adjusted mode of 24 events in the range of $100,000–$250,000, which is somewhat less than the typical current cost of a home (although prices are quickly rising). Such a value indicates that a home was not totally destroyed. The largest home damage is in the category of $750,000 to one million dollars.
- **Buildings and objects** (Table 5.28): They range from many small losses, typically trees, to very expensive damages.
- **Top ten loss events** (Table 5.29): These are most often structures other than homes.

The most expensive event of 116 million dollars in the dataset (Table 5.29) is a 1990 forest fire caused by lightning in the Tonto National Forest in Gila County that destroyed 28 million board feet of trees over 28,500 acres. At the time, it was the worst wildfire in Arizona history. The estimate provided is taken from web-based prices for lumber, but there are many variables, and it may not be especially accurate. Nevertheless, there have been additional large forest fire events destroying lumber before and since 1990, but this is the only forest fire event included in *Storm Data*.

5.3.8 Insurance Losses

According to the Insurance Information Institute, there were 76,860 insurance claims for lightning losses in the U.S. in 2019, for a cost to insurers of $920 million. Many of these were in populous states with large lightning frequencies, including Florida. The average claim was nearly $12,000, and these costs are steadily rising, mainly due to more electronics in homes. This is a major increase since the study mentioned earlier (Holle et al. 1996) that found an average of $873 for personal claims and $1,369 for commercial claims 30 years ago.

What about Arizona? Insurance data for the state in 2019 were kindly provided by Ms. Loretta Worters at the Insurance Information Institute. Arizona ranked number 29 in the U.S. in the number of homeowner lightning insurance claims, which is similar to the state ranking of number 33 in the number of CG flashes (Sect. 4.2.4). There were 1,020 homeowner claims reported to the Institute in 2019, at an average cost of $10,725 per claim. The resulting total value of claims in Arizona for 2019 was $10,900,000.

5.3.9   Forest Fires

Large areas of Arizona are forested. There are six national forests across the state and stands of trees are located on all of the mountains at higher elevations as shown in the map in Fig. 2.1. Forests cover huge swaths of land on the north and south rims of the Grand Canyon, along the Mogollon Rim across central Arizona, in the White Mountains in the east-central part of the state, and in the sky islands of southern Arizona. Since Arizona is also a relatively dry state subject to intermittent dry and wet years, it is not surprising that forest fires are common.

Dry lightning refers to CG flashes occurring when little or no rain is falling on the ground. This situation is frequent throughout the Western states and leads to fires starting when the fuel, moisture content of the air and ground, and other factors are in a dangerous combination. Forest fires started by lightning in Fig. 5.9 from 2000 through 2019 are based on data at https://gacc.nifc.gov/swcc/predictive/intelligence/Historical/Fire_and_Res ource_Data/Historical_Fires_Acres.htm. There are two main results: (1) a large number of forest fires are started by lightning every year in Arizona, and (2) the number is variable from year to year. For example, the number of lightning-caused fires in Fig. 5.9a shows a continued decrease. However, it is somewhat suspicious that some types of reporting have changed—the number drops abruptly from 2006 to 2008 and stays smaller. Nevertheless, there are hundreds of lightning-caused fires every year in Arizona. Figure 5.9b shows that the lightning-related percentage of all forest fires reduces from over 60% to 20% in later years, subject to the same concern about a change in reporting. One of many possible causes of such a change is whether the fires were human-caused or by lightning; these are difficult assessments to make. The acreage burned by lightning-caused forest fires is often between 100,000 and 200,000 acres (Fig. 5.9c). Although the lightning-caused fires are not

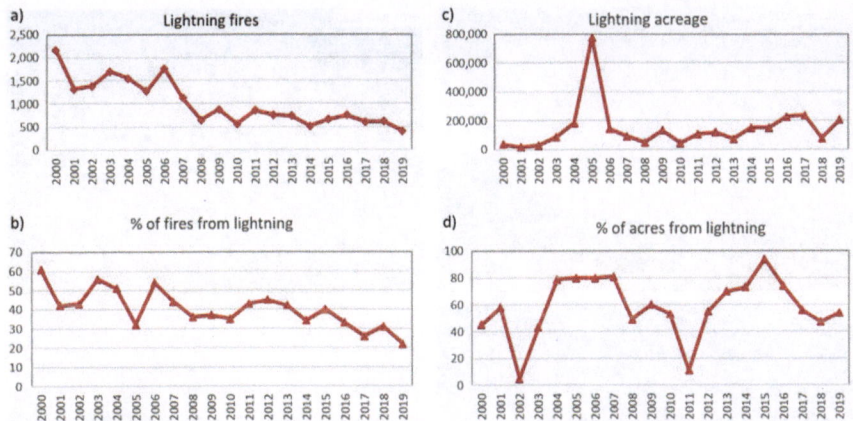

Fig. 5.9 a Number of Arizona forest fires caused by lightning, b % due to lightning, c acreage of lightning fires, and d % of acres from lightning from 2000 to 2019

much more than half of the events, they burn 80% of the acreage (Fig. 5.9d) over many years.

On June 5, 2020, a dry lightning event over the Santa Catalina Mountains near Tucson began a fire that lasted for nearly two months and unfortunately destroyed much of the tree and other vegetation cover in the Coronado National Forest on top of the Catalinas. Figure 5.10 shows a series of photos from the first author's home during the course of June. Active fire could be seen on the mountain every night for weeks and smoke was visible during the day for all of this period. In this case, there was not much lightning, on the order of a few dozen CG flashes. However, as shown in Sect. 4.3.2, lightning is uncommon in June, especially in early June. Nevertheless, this small lightning outbreak caused lightning (1) only over mountains, (2) that had trees and large bushes and ground cover, (3) after several months of the usual dry springtime weather, and (4) under very dry and windy conditions on a specific day in early June. The fire spread relentlessly in all directions from the initiating point that was well located by the National Lightning Detection Network (NLDN) described in the following chapter. Although no people were injured and no buildings were damaged, the Catalinas will show scars from this fire for many decades into the future.

Fig. 5.10 Photos of Bighorn Fire as viewed looking east from Oro Valley, Arizona. Panels **a**, **b**, and **c** show the spread of the fire during the first day after it began on the evening of June 5, 2020. Subsequent panels show its spread and persistence for weeks afterward (Panels a, b, c, d, f, © R. Holle; panel e © S. Holle)

Pyrocumulus

Figure 5.10e and f show clouds billowing vertically from the Bighorn Fire. These clouds are composed mainly of smoke, but sometimes they can ingest water droplets, ice particles, and ice crystals that are in a normal thunderstorm. In this case, the storm may follow the sequence of thunderstorm development as in Sect. 1.3, resulting in lightning that compounds the difficulty of coping with a large forest fire.

Lauridsen

On a lighter but relevant note, both authors of this book have a life-long interest in continuously learning about classical music, and the second author has composed a number of selections from a classical approach. Morton Lauridsen is one of the leading living composers of our time and received the National Medal of Arts in 2007. A one-on-one conversation was held with Dr. Lauridsen at a concert including his music by the True Concord Chorus and Orchestra in Tucson in March 2019. He recounted how he worked as a fire lookout near Mt. St. Helens in Oregon in 1969. He spent several weeks nonstop at the top of a lookout tower and was tasked with spotting lightning and subsequent lightning-caused fires that sometimes flared up several days later. As lightning struck all around the tower, he sat on an insulated chair in the middle of the room at the top inside the tower; several of his colleagues were injured by lightning that summer. It's not too surprising, then, that this aspiring world-famous composer said he spent parts of days and nights playing his trumpet with coyotes howling in the distance. Since then, the lightning detection technology developed at the UArizona described in Chap. 6 has supplanted the need for nearly all fire lookout towers in the U.S.

5.3.10 Fulgurites

No discussion of lightning impacts is complete without mentioning fulgurites (Pasek et al. 2012). While they are not often found in Arizona, fulgurites are available at the Tucson Gem and Mineral Show in February. They are formed when lightning strikes sandy soil and turns the grains into rock, usually in long tubes. There are a few small areas of sand dunes in Arizona along the Yuma border area and on the Navajo Nation, contrary to popular opinion from out of state. Nevertheless, the first author decided to have a special license plate made in their honor (Fig. 5.11). What could be more fitting than a license plate with a lightning phenomenon on it, in Arizona, and with a UArizona logo where so much lightning research has taken place (Chap. 7)? It's very likely to be the only plate in the world with FULGUR on it!

Extreme Fulgurite

Normally, fulgurites are a few inches long. However, the longest example to date was 14 feet long, excavated in north Florida. It was found by a group headed by Dr. Martin Uman, one of the co-inventors of real-time lightning detection while at the UArizona (Chap. 6). The Florida fulgurite was in sandy soil, about an inch in diameter, mostly vertical, and had multiple branches (Wright 1998). This is more than an academic exercise since underground

Fig. 5.11 Fulgurite license plate (© R. Holle)

power lines are susceptible to lightning damage due to processes similar to those that form fulgurites.

5.4 Positive Roles of Lightning in Atmospheric NOx Cycle and Global Electric Circuit

We have discussed the negative effects of lightning extensively in this chapter, but we want to emphasize and end with the positive effects of lightning.

Since lightning kills people and causes damage, would it be better if there was no lightning? The answer is no. Although dangerous, lightning has a large number of beneficial effects for us. First of all, lightning may have played a vital role in the origin of life on Earth. As early as 1871, Charles Darwin put forward the idea of lightning sparked the original life as follows:

> …we could conceive in some warm little pond, with all sorts of ammonia and phosphoric salts, light, heat, electricity, &c., present, that a protein compound was chemically formed ready to undergo still more complex changes, at the present day such matter would be instantly devoured or absorbed, which would not have been the case before living creatures were formed (Darwin 1871).

A recent study (Hess et al. 2021) suggests that lightning might have served as a source of prebiotic, an element that potentially facilitates the emergence of habitable environments.

Lightning helps fertilize the soil by producing nitrate nutrients in the atmospheric nitrogen cycle. Nitrogen makes up 78% of the atmosphere, and a nitrogen molecule consists of two atoms which are tightly bonded together. The energy released by lightning strikes can separate them as single nitrogen atoms. The nitrogen atoms can easily bond to oxygen in the atmosphere to form nitrogen oxides (NOx). Once formed, nitrogen oxides can be carried by the rain droplets to the ground that can be absorbed by plants. People, therefore, obtain nitrogen from eating these plants to help maintain the health of our bodies. Lightning may also serve as an activator for cleaning our atmosphere, according to a recent study by Brune et al. (2021). Lightning generates 2–16% of the global hydroxyl radical (OH), which is one of the most reactive gases in the atmosphere and is of significance in controlling toxic gases like carbon monoxide and climate-relevant gases like methane (Li et al. 2018).

In addition, lightning plays a key role in the cycle of the Global Electric Circuit (GEC). The GEC is a system between the earth and ionosphere that produces a continuous electric field of roughly 100 to 300 V per meter (Fig. 5.12). Thunderstorms also generate currents to maintain the cycle by transferring charge through lightning (Wilson 1920). Studies have shown a close relationship between GEC and global variations in the multiple-year cycle called the El Niño/Southern Oscillation (Slyunyaev et al. 2021).

Lastly, let's remember that by itself, lightning is astonishing to watch and appreciate its beauty, especially if it is seen with the unique Arizona landscape and background, as described in detail in Chap. 3. As also shown in Chap. 3,

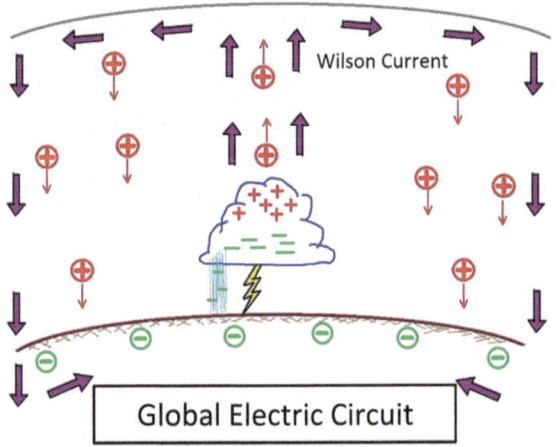

Fig. 5.12 Illustration of the Global Electric Circuit and Wilson Current. Courtesy of Dr. Charles Weidman at the UArizona

lightning is one of the most amazing atmospheric phenomena that attracts professional and amateur photographers to observe, chase, capture, and enjoy. But as the present chapter has shown, it also is dangerous and demands a great deal of respect and caution!

References

Anderson RB (2001) Does a fifth mechanism exist to explain lightning injuries? IEEE Eng Med Biol Mag 20(1):105–113

Anderson RB, Jandrell I, Nematswerani H (2002) The upward streamer mechanism versus step potentials as a cause of injuries from close lightning discharges. Trans SA Inst Elec Eng 93(1):33–43

Anderson RB, Carte AE (2009) Struck by lightning. Archimedes 09:25–29

Andrews CJ (1992) Telephone related lightning injury. Med J Aust 157(11/12):823–825

Andrews CJ, Darveniza M (1989) Telephone mediated lightning injury—an Australian survey. J Trauma 29(5):665–671

Andrews CJ (2021) A study of earth potential rise shock in lightning injury. In: Preprints of the 35th international conference on lightning protection and international symposium on lightning protection (XVI SIPDA), Colombo, Sri Lanka, 20–26 Sept 2021

Blumenthal R (2012) Secondary missile injury from lightning strike. Am J for Med Pathol 33:83–85

Blumenthal R, West NJ (2015) Investigating the risk of lightning's pressure blast wave. S Afri J Sci 111(3/4)

Brune WH, McFarland PJ, Bruning E et al (2021) Extreme oxidant amounts produced by lightning in storm clouds. Science 372(6543):711–715

Cherington M, Walker J, Boyson M et al (1999) Closing the gap on the actual numbers of lightning casualties and deaths. In: Preprints of the 11th conference on applied climatology. American Meteorological Society, Dallas, Texas, 10–15 Jan 1999

Cooper MA (2002) A fifth mechanism of lightning injury. Acad Emerg Med 9(2):172–174

Cooper MA, Holle RL (2018) Reducing lightning injuries worldwide. Springer Natural Hazards, New York, p 233

Cooper MA, Holle R, López R (1999) Recommendations for lightning safety. J Am Med Assoc 282:1132–1133

Cooper MA, Marshburn S Sr, Marshburn J (2002) Lightning strike and electric shock survivors. Int Nat Wea Digest 25:48–50

Cooper MA, Holle RL (2007) Casualties from lightning involving motorcycles. In: Preprints of the international conference on lightning and static electricity, Paris, France, 28–31 Aug 2007

Cooper MA, Holle, RL (2011) Mechanisms of lightning injury should affect lightning safety messages. Newsletter, Nat Wea Assoc, June

Cooper MA, Holle RL, Andrews CJ (2008) Distribution of lightning injury mechanisms. In: Preprints of the 2nd international lightning meteorology conference, Vaisala, Tucson, Arizona, 21–23 April 2008

Cooper MA, Blumenthal R, Silva LM et al (2015) A follow-up study of a large group of children struck by lightning. In: Preprints of the 7th conference on the meteorological applications of lightning data. American Meteorological Society, Phoenix, Arizona, 4–8 Jan 2015

Cooper MA, Gomes C, Tushemereirwe R et al (2016) The development of the African centres for lightning and electromagnetics network. In: Preprints of the 33rd international conference on lightning protection, Estoril, Portugal, 25–30 Sept 2016

Cooper MA, Andrews CJ, Holle RL et al (2017) Lightning-related injuries and safety. Chapter 5, Wilderness Medicine, 7th Edition. Elsevier, Philadelphia, Pennsylvania, P. Auerbach, Editor

Cooper MA, Tushemereirwe R, Holle RL (2018) African Centres for Lightning and Electromagnetics Network (ACLENet): progress report. In: Preprints of the 6th international lightning meteorology conference, Fort Lauderdale, Florida, 12–15 March 2018

Cooper MA, Holle RL, Tushemereirwe R et al (2019) African centres for lightning and electromagnetics network (ACLENet): application to South America? In: Preprints of the international symposium on lightning protection (XV SIPDA), Sao Paulo, Brazil, 30 Sept–04 Oct2019

Cummins KL, Krider EP, Olbinski M et al (2018) A case study of lightning attachment to flat ground showing multiple unconnected upward leaders. Atmos Res 202:169–174

Curran EB, Holle RL, López RE (2000) Lightning casualties and damages in the United States from 1959 to 1994. J Clim 13:3448–3453

Darwin C (1871) "Letter to Joseph Hooker". In: The life and letters of Charles Darwin, Including an autobiographical chapter, 3:18, John Murray

De Fonvielle W (1869) Thunder and lightning. Charles Scribner & Co., New York

Elsom DM (2015) Lightning: nature and culture. Reaktion Books, London, U.K., p 240

Florida Division of Emergency Management (2004) The adventures of rabbit, possum, and squirrel in the 30/30 rule. Tallahassee, Florida. https://training. fema.gov/hiedu/aemrc/courses/coursesunderdev/crisisandrisk/crisis-risk%20c ommunications%20-%20session%2011%20-%20handout%2011-2%20adve ntures%20of%20squirrel%20rabbit%20possum.pdf

Ghiglieri MP, Myers TM (2001) Over the edge: death in grand canyon. Puma Press, Flagstaff, Arizona, p 408

Gomes C, Izadi M (2019) Lightning caused multiple deaths: Lethality of taking shelter in unprotected buildings. In: Preprints of the international symposium on lightning protection (XV SIPDA), Sao Paulo, Brazil, 30 Sept–04 Oct 2019

Gonzales L (2004) Deep survival: who lives, who dies, and why. Norton and Company, Inc., 304 pp

Hess BL, Piazolo S, Harvey J (2021) Lightning strikes as a major facilitator of prebiotic phosphorus reduction on early Earth. Nat Commun 12(1):1–8

Holle RL (2014) Diurnal variations of NLDN-reported cloud-to-ground lightning in the United States. Mon Wea Rev 142:1037–1052

Holle RL (2016) A summary of recent national-scale lightning fatality studies. Wea Clim Soc 8:35–42

Holle RL, López RE, Arnold LJ et al (1996) Insured lightning-caused property damage in three western states. J Appl Meteor 35:1344–1351

Holle RL, López RE, Zimmermann C (1999) Updated recommendations for lightning safety-1998. Bull Am Meteor Soc 80:2035–2041

Holle RL, López RE, Navarro BC (2005) Deaths, injuries, and damages from lightning in the United States in the 1890s in comparison with the 1990s. J Appl Meteor 44:1563–1573

Holle RL, Demetriades NWS, Nag A (2016) Objective airport warnings over small areas using NLDN cloud and cloud-to-ground lightning data. Wea Forecast 31:1061–1069

Holle RL, Brooks WA, Cummins KL (2021) Lightning occurrence and casualties in U.S national parks. Wea Clim Soc 13:525–540

Holle RL, Cooper M (2019) Overview of lightning injuries around the world. In: Preprints of the 11th Asia-Pacific international conference on lightning, Hong Kong, China, 12–14 June 2019

Holle RL, Krider EP (2006) Suspension of a University of Arizona football game due to lightning. In: Preprints of the 19th international lightning detection conference, Vaisala, Tucson, Arizona, 24–25 April 2006

Holle RL, Zhang D (2017) So you think you know lightning: a collection of electrifying fast facts. Vaisala, Inc., 64 pp. https://www.vaisala.com/en/system/files/documents/Lightning-Booklet.pdf and https://lightningdev.umd.edu/aert/Safety.html

Holle RL (2005) Lightning-caused recreation deaths and injuries. In: Preprints of the 14th symposium on education, American Meteorological Society, San Diego, 9–13 Jan 2005

Holle RL (2008a) Annual rates of lightning fatalities by country. In: Preprints of the 20th international lightning detection conference, Vaisala, Tucson, Arizona, 21–23 April 2008a

Holle RL (2008b) Lightning-caused deaths and injuries in the vicinity of vehicles. In: Preprints of the 3rd conference on meteorological applications of lightning data, American Meteorological Society, New Orleans, Louisiana, 20–24 Jan 2008b

Holle RL (2010) Lightning-caused casualties in and near dwellings and other buildings. In: Preprints of the 3rd international lightning meteorology conference, Vaisala, Orlando, Florida, 21–22 April 2010

Holle RL (2012) Lightning-caused deaths and injuries in the vicinity of trees. In: Preprints of the 31st international conference on lightning protection, Vienna, Austria, 2–7 Sept 2012

Jensenius JS, Franklin DB (2014) NOAA's efforts to reduce lightning fatalities through public education and awareness. In: Preprints of the 5th international lightning meteorology conference, Vaisala, Tucson, Arizona, 20–21 March 2014

Jensenius JS (2016) A detailed analysis of lightning deaths in the United States from 2006 through 2015. In: Preprints of the 14th symposium on education, American Meteorological Society, San Diego, California, 9–13 Jan 2005

Li M, Karu E, Brenninkmeijer C et al (2018) Tropospheric OH and stratospheric OH and Cl concentrations determined from CH_4, CH_3Cl, and SF6 measurements. NPJ Clim Atmos Sci 1(1):1–7

López RE, Holle RL (1996) Fluctuations of lightning casualties in the United States: 1959–1990. J Clim 9:608–615

López RE, Holle RL (1998) Changes in the number of lightning deaths in the United States during the twentieth century. J Clim 11:2070–2077

López RE, Holle RL, Heitkamp TA et al (1993) The underreporting of lightning injuries and deaths in Colorado. Bull Am Meteor Soc 74:2171–2178

Mäkelä J, Karvinen E, Porjo N et al (2003) Attachment of natural lightning flashes to trees: preliminary statistical characteristics. J Lightning Res 1:9–21

Noggle C, Byerley L, Harlan TP et al (2002) Magnetic evidence of lightning currents on the ground. In: Preprints of the international lightning detection conference, Vaisala, Tucson, Arizona, 16–18 Oct 2002

Pasek MA, Block K, Pasek V (2012) Fulgurite morphology: a classification scheme and clues to formation. Contrib Mineral Petrol 164:477–492

Raga GB, de la Para MG, Kucienska B (2014) Deaths by lightning in Mexico (1979–2011): threat or vulnerability? Wea Clim Soc 6:434–444

Rakov VA, Uman MA (2003) Lightning: physics and effects. Cambridge University Press, New York, p 683

Roeder WP, Holle RL, Cooper MA et al (2012) Lessons learned in communicating lightning safety effectively. In: Preprints of the 4th international lightning meteorology conference, Vaisala, Broomfield, Colorado, 4–5 April 2012

Roeder WP (2009) Research required to improve lightning safety. In: Preprints of the 4th conference on the meteorological applications of lightning data, Phoenix, Arizona, American Meteorological Society, Phoenix, Arizona, 11–15 Jan 2009

Roeder WP (2012) Lightning has fallen to third leading source of U.S. storm deaths. In: Preprints of the annual meeting, National Weather Association, Madison, Wisconsin, 06–11 Oct 2012

Roeder WP (2014) Backcountry lightning risk reduction—Lightning crouch versus standing with feet together. In: Preprints of the 5th international lightning meteorology conference, Vaisala, Tucson, Arizona, 20–21 March 2014

Roeder WP (2016) Changes in U.S. annual lightning fatalities from 1990–2015. In: Preprints of the 6th international lightning meteorology conference, Vaisala, San Diego, California, 18–21 April 2016

Schonland BFJ (1950) The flight of thunderbolts, 2nd edn. Clarendon Press, Oxford

Slyunyaev NN, Ilin NV, Mareev EA et al (2021) The global electric circuit land–ocean response to the El Niño—Southern Oscillation. Atmos Res 260:105626

Steenbergh WF (1972) Lightning-caused destruction in a desert plant community. Southwest Nat 16(3/4):419–429

Steiner M, Deierling W, Bass R (2013) Balancing safety and efficiency of airport operations under lightning threats: a look inside the ramp closure decision-making process. J Air Traffic Control 55:16–23

Steiner M, Deierling W, Ikeda K et al (2016) Ground delays from lightning ramp closures and decision uncertainties. Air Traffic Control Quart 22:223–249

Tomlinson C (1848) The Thunders-storm, or, an account of the nature, properties, dangers, and uses of lightning in various parts of the world. Society for Promoting Christian Knowledge, London

Uman MA (1986) All about lightning. Dover Press, 167 pp

Walsh KM, Cooper MA, Holle R et al (2013) National athletic trainers' association position statement: lightning safety for athletics and recreation. J Athletic Training 48:258–270

Wilson CTR (1920) Investigations on lightning discharges and on the electric field of thunderstorms. Philos Trans Roy Soc Lond Ser A 221:73–115

Wright FW (1998) Florida's fantastic fulgurite find. Weatherwise 51:28–31

Yas V, Sánchez Núñez P, Trujillo-Falcón JE et al (2021) When thunder roars, go indoors? Spanish-language resources for national lightning safety awareness week. In: Virtual presentation at the 101st annual meeting of the American Meteorological Society, 10–15 Jan 2021

Zhang DL (2022) The legacy of the Chinese lightning and thunder gods. Weatherwise 75:24–28

6

How Lightning Detection Networks Were Developed in Arizona

Abstract Real-time lightning detection over large regions began in the middle 1970s at the University of Arizona. It was based on a foundation of basic research there and elsewhere. But it was also the combination of prior knowledge, advances in integrated circuits, communications, and computer capabilities that were necessary to bring it to fruition. In addition, it should be pointed out that basic research by faculty and students at universities, interest from private industry, and government support all contributed and combined to make this happen. The systems were an immediate success, resulting in a growing number of applications and customers such as the first users for forest fire detection and electric utility needs. Companies were formed and grew, and Tucson became a hub for this development.

6.1 Invention of Real-time lightning Detection at the University of Arizona

Why would anyone study purely basic principles of lightning in 1970? What good could it do? Isn't open-ended basic research such as this only interesting to a few people? Curiosity-based lightning research was not progressing especially quickly around that time, but that situation was about to change within a few short years.

What if it turns out that the results of these basic studies changed a whole segment of the scientific world? What if it saved lives, reduced damages, injuries, and deaths, and made society operate more safely and efficiently?

© Springer Nature Switzerland AG 2023
R. L. Holle and D. Zhang, *Flashes of Brilliance*,
https://doi.org/10.1007/978-3-031-19879-3_6

How do you measure the value of the seed money that starts such a process? Who decides what will turn out to be a great idea and which will not?

Of course, you can't tell for sure in advance, but the indications were there by the middle 1970s that such an invention was indeed developing at the University of Arizona (UArizona) in lightning. This was not serendipity when something happens by chance or luck. Instead, the foundation was built on the prior knowledge, capabilities, insight, and technological awareness of a small number of people. We'll try to tell the story in this chapter in some detail but keep the main points in mind.

Question

Why was the development of real-time lightning detection at the UArizona in the 1970s not due to serendipity? How else can it be explained?

So how did this happen? It was a combination of the right people at the right place at the right time. Lightning-related basic research and a variety of talents connected with the latest advances in electronics, computers, and communications. Moreover, it was a combination of academia, private industry, and government that carried it forward. Without all these factors, the real-time private industry of lightning detection would not have happened when it did. In addition, scientific advances are built on the foundations provided by earlier efforts, so it takes time to reach what can seem to be an isolated breakthrough at a specific time. Instead, the seemingly obscure and slow process of basic research was necessary to make a large leap forward to a new subdiscipline in science, that of precise real-time lightning detection.

What exactly is the accomplishment? If you turn on a local television news broadcast, or a national cable weather channel, or use a cell phone or desktop weather provider, you see real-time lightning data. The information varies by version, quality, display, and coverage, but nearly every thunderstorm in the world is detected by one or more of these networks. In addition, most individual flashes are detected, and these are not only located precisely in place and time, but their intensity, polarity, and quality of measurement is known, as well as whether they came in contact with the ground or stayed in cloud.

6.2   Overview of Lightning Detection Milestones

Before going more deeply into the details, let's list the main accomplishments as reference points for the following sections:

- 1970s

  - The middle 1970s: Basic research at the UArizona builds on in-house and external developments in the global lightning community that leads to real-time lightning detection.
  - The late 1970s: First single-station sensors deployed in Alaska for forest fire detection.
  - The late 1970s: First multiple-station networks in Alaska and parts of western U.S. for forest fire detection, and local networks in Florida for research.
  - The late 1970s: Formation of Lightning Protection and Location Inc. (LLP) in Tucson.
  - The late 1970s: First International Lightning Detection Conference (ILDC) in Tucson.

- 1980s

  - The early 1980s: Network across the western U.S. deployed for forest fire detection by the Bureau of Land Management (BLM).
  - The 1980s: Installation of multi-station networks begins in several countries around the world.
  - The early to middle 1980s: Establishment of regional networks by the State University of New York at Albany (SUNYA), National Severe Storms Laboratory (NSSL), and BLM for growing interests in research, applications, and utility interests.
  - The late 1980s: Formation of Global Atmospherics Inc. (GAI) in Tucson.
  - 1989: Dr. Ken Cummins and Patrick Zumbusch arrive at GAI.

- 1990 forward

  - The 1990s: Emphasis on improving location accuracy and detection efficiency of the National Lightning Detection Network (NLDN).
  - 1993: Network Control Center established in Tucson after relocating from SUNYA.
  - 1993: GeoMet Data Services (GDS)/LLP acquires Atmospheric Research Systems, Inc. (ARSI), then combines direction finding and

time of arrival technology into the IMProved Accuracy from Combined Technology (IMPACT) sensor.

- 1994: NLDN upgrade with IMPACT sensors
- 1998: Canadian Lightning Detection Network (CLDN) established.
- 2002: GAI acquired by Vaisala Inc.
- The early 2000s: Addition of in-cloud lightning detection.
- The 2000s: The EUropean Cooperation for LIghtning Detection (EUCLID) network is formed to provide seamless lightning detection among most countries in western Europe.
- The late 2000s: Beginning of efforts to detect long-range lightning outside U.S.
- The early 2010s: Formation of the Global Lightning Dataset (GLD360) network covering the world.
- 2019: Network Operations Center moves from Tucson to Louisville, Colorado.

6.3   Lightning Detection Development at the University of Arizona

In the years prior to the development of lightning detection at Arizona, Dr. Leon Salanave had been studying the visible light spectrum of lightning. At that time, lightning was not nearly as widespread a field of study as it is today, so it was difficult to develop a collection of lightning-related publications from around the world. Dr. Salanave persisted in doing so in the 1960s and compiled a notebook with journal articles and other available resources into what he called 'Der Blitz Bibel' (see more in Sect. 7.1.1). This collection was provided to us by Dr. E. Philip Krider from the Institute of Atmospheric Physics at the UArizona, and we have found additional useful resources in this partially hand-written documentation. The lightning studies first shown in Salanave (1961) culminated in his 1980 book *Lightning and Its Spectrum* (Salanave 1980) containing amazing lightning photos described in Sect. 3.2.2. A historical perspective on the importance of these lightning spectroscopy studies was published by Walker and Christian (2014). Before the book was published, Dr. Salanave had left UArizona for the position of staff astronomer at the Morrison Planetarium of the California Academy of Science. All but one reference in the 1980 book is from 1977 or earlier, so the idea of real-time lightning detection was not even mentioned!

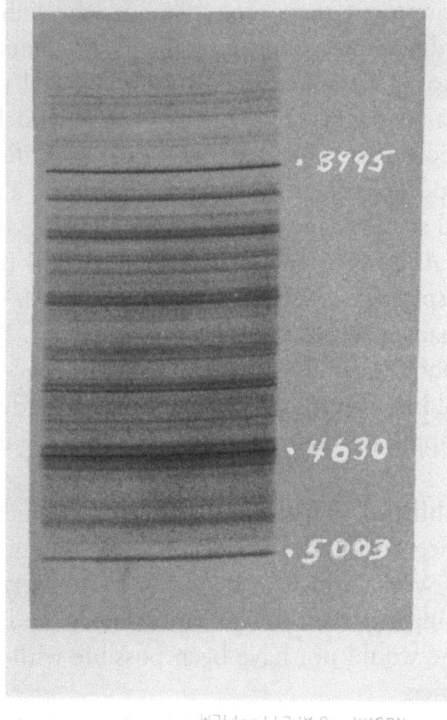

Fig. 6.1 Wavelengths of a lightning channel photographed using slit spectrography from Lowell Observatory in Flagstaff on July 24, 1917

An interesting historical background was found while searching through Der Blitz Bibel. Slipher (1917) at the Lowell Observatory in Flagstaff performed very early studies of the chemical elements within the spectrum of a lightning channel. Figure 6.1 shows a negative of a photo taken there on July 24, 1917, with a slit spectrograph that is used in astronomy. The thunderstorm was on the south slope of the San Francisco Peaks, about nine miles north of the Observatory. The numbers label the wavelengths of major specific atomic bands shown in the photo, and from the text of Slipher (1917), it is apparent that this is among the first such images taken in the world and is one of the first lightning-related photos taken in Arizona, although not of an actual lightning channel. Since this was ground-breaking research and took place within Arizona, Salanave included it in his background studies and somehow was able to obtain the image! A photo of lightning occurring over the Grand Canyon taken by the Kolb brothers at about the same time in the early 1910s is shown in Fig. 3.3.

A dramatic development then took place in the lightning community. A core group of individuals with complementary talents had assembled at the UArizona in mostly basic research. In alphabetical order, they are Dr. E. Philip Krider, Carl Noggle, Dr. Richard Orville, and Dr. Martin Uman. Through collaboration over about a decade, they identified the key theoretical and practical aspects of what it would take to build a real-time lightning detector, link several sensors together to locate lightning accurately, and send such data to users. Additional core individuals entered the development of the U.S. NLDN, especially Dr. Ken Cummins of GAI/Vaisala and then at the university. We had the privilege of interviewing Drs. Krider, Orville, and Cummins for this book (Sect. 7.4).

Key technological innovations during this time took advantage of the latest advances that were applied to the discipline of improving lightning detection:

- **Computer capabilities**: The rapid rise in computational speed and memory took place during the period of development of real-time lightning detection networks from the late 1970s to the early 2000s. The ability to process incoming signals from multiple sensors and multiple lightning events in real time would not have been possible without this spectacular well-known advance.
- **Integrated circuits [Cummins interview, Sect. 7.4.2]**: "What got LLP started and able to make a commercial sensor around 1980 was the development of integrated circuits that put more than just a transistor or a couple of transistors into a package. You needed to have good operational amplifiers that can do good analog signal conditioning, and you can build them in circuits fairly easily. They could make analog circuits that could amplify things and make a good signal. And then you needed to have to convert it to digital and then do a lot of processing with that. And to do that, you had to have microcontrollers, and both of them grew up together from the late 1970s to the mid-1980s. Without those, you would have never had the first model 80-02 sensor."
- **Communications [Krider interview, Sect. 7.4.1]**: "Communication technology in general was pretty poor. We had trouble getting a 300 baud phone line! Carl (Noggle) had developed a way of using the horizontal sync pulse video signal. We were experimenting with ways of getting timing. And one way of getting good multiple-station timing was within the range of a television transmitter and syncing on the horizontal sync pulse of the video signal. Those pulses come very very precisely, and you can pick them up anywhere that you can pick up a television signal. And we were experimenting with synchronizing multiple stations around the Tucson area

using those sync pulses. That was before the GPS receivers came out in a very compact cheap format. You could buy a GPS receiver starting in about the early or mid-80s for $10,000, but clearly that's too expensive to run on a lightning sensor. So, we were experimenting with other ways of getting multiple stations synchronizing. And so, Carl had synchronized a television display to the output of a single-station direction finder, so we could plot what the television saw at the same time we plotted the vectors. And when you looked at the display of the magnetic direction on top of the image, you can really tell what it is that's making the pulse. And so, we can see what kind of pulses made the return strokes and what kind of pulses made other things, and the other things you didn't want obviously depending on a lot of what the source was—if it was a Xerox machine, which was a frequent problem. Xerox machines make pulses that look exactly like return strokes!"

- **Communications [Cummins interview, Sect. 7.4.2]**: "And that's the initial development of GPS, a few of the satellites got put up, they made these giant GPS receivers that were the sizes of a loaf of bread or a half a loaf of bread. They're extremely expensive, but we knew that they were going to get cheaper…GPS, we knew that there was such a pull and a demand for it that not only was it going to be technically adequate to get the timing that you want, sub-microsecond or microsecond level timing, but that it was going to become cheap in a short amount of time."

Three critical patents were issued during the early stages of this development. They were needed to protect the theoretical and technological breakthroughs involved in real-time lightning detection. They are:

- Patent 4,115,732 by Krider, Noggle, and Uman with the assignee of the University of Arizona Foundation titled "Detection system for lightning". The application was filed on 14 October 1976 and issued on 19 September 1978.
- Patent 4,198,599 by Krider and Noggle with the assignee of the University of Arizona Foundation titled "Gated lightning detection system." The application was filed on 31 July 1978 and issued on 15 April 1980.
- Patent 4,245,190 by Krider, Noggle, and Uman was assigned to Lightning Location and Protection, Inc. titled "Lightning detection system utilizing triangulation and field amplitude comparison techniques". The application was filed on 12 December 1978 and issued on 13 January 1981.

Patent

If you have never read a patent, take a look at one of them. They are examples of how a breakthrough device or process must be documented in complete detail to be protected. They are not trivial!

These concepts were described in journal articles by Krider and Noggle (1974), Krider et al. (1976), Herman et al. (1976), Lin et al. (1979), Krider et al. (1980), and Uman et al. (1980). These foundational papers on the topic of lightning detection originated from the UArizona. It is apparent that publication in formal peer-reviewed mainstream journals was considered an important part of the process from the start of the development of lightning detection, in order to establish the sound scientific basis of the technology. In addition, this is an unusual instance in science…there is a beginning to a discipline. We can trace its history back to these papers, since there were none related to functioning real-time ground-based networks before 1976. Of course, the technique is based on knowledge and expertise prior to this time, as described in Chap. 7. Note that a current bibliography of publications using all types of ground-based real-time lightning detection methods around the world is now approaching 4,000 papers.

Dr. Richard Orville brought forward the sensor technology breakthrough at the UArizona when he became an Atmospheric Science professor at SUNYA. Through an arrangement with the Electric Power Research Institute (EPRI), an eastern U.S. network was installed (Orville et al. 1983a, b; Orville 2008). EPRI is the research arm of more than half of the U.S. electric utilities and saw the benefits of real-time lightning detection at an early stage. The first SUNYA sensors installed in 1982 became part of a ten-station network in 1983 from New York to Virginia displaying locations using the method of Orville Jr. (1987). Ron Henderson and others established an Operations Center in Albany using a large number of linked personal computers. By 1989, SUNYA completed its network across most of the U.S. when it complemented networks operated by NOAA's NSSL on the southern Plains, and BLM in western states. Applications of the SUNYA network data developed quickly, such as at the National Severe Storms Forecast Center of the National Weather Service (Lewis 1989), Tennessee Valley Authority (Whitehead and Driggans 1990), Consolidated Edison of New York (Idone and Orville 1990), and Detroit Edison Company (Whitney and Asgeirsson 1991).

In Arizona, the developments were often highlighted in the media because this technology was big news. Here are selected articles and benchmark events from Arizona and other locations in the earlier years:

- *Tucson Daily Citizen* newspaper, 09 August 1973: Describes the ongoing research by Krider and Noggle at the UArizona, with Uman at the University of Florida, as they were developing a lightning detection system in reaction to the 1969 Apollo 12 lightning strike. Krider is described as "working on a photoelectric lightning detector..." The antenna design shown in the article is a precursor to the first network antennas used a few years later.
- *Journal of Spacecraft and Rockets*, 1974: Detailed analysis of the plume from a rocket launch is made for the first time from the perspective of how lightning interacts to initiate a flash (Krider et al. 1974).
- *The Dryden Observer* newspaper, Ontario, Canada, 02 August 1978: A spokesman for Canada's first lightning detection system stated that "if this equipment catches one fire before it reaches the project stage...it's very cost effective."
- *Tucson Daily Citizen* newspaper, 14 June 1979: Lightning detectors in Alaska are being used where fire control authorities often send crews to fight a fire, no longer relying on a special detection plane. The Canadian government was considering a similar system.
- *Tucson Citizen* newspaper, 28 September 1979: "Lightning detector battles fires" describes how BLM is saving millions of dollars by reducing the effort to find lightning-caused fires in the West and Alaska. In fact, the first year of deployment in Alaska is said to have paid for itself within two weeks.
- *The Arizona Daily Star* newspaper, Tucson, 30 January 1981: "Their 'simple breadbox' spots lightning in a flash" describes their presentations at the ILDC about the first lightning location system introduced in 1976. By 1981, 50 sensors were installed to cover 40% of the U.S., as well as networks in Norway, Sweden, Australia, and South Africa.
- *Maclean's magazine*, 18 April 1983: Alberta, Canada's forest services is planning a summer trial of the new lightning detection system that "could provide the first real breakthrough in the campaign to ease an awesomely costly annual disaster".
- *The Economist*, 13 August 1983: The BLM system is being expanded in the western states so that "Help is on the way for the rangers perched in isolated watchtowers trying to spot blazes before they get out of control."

- *The Arizona Daily Star* newspaper, 19 January 1984: Reports on the attendance by 85 people from around the world at ILDC in Tucson to exchange information about the "revolutionary" lightning detection systems.
- National Weather Association Operational Achievement Group Award, 1985: The Automatic Lightning Detection System team of BLM was recognized for operating the BLM network of LLP direction finder sensors in the western U.S.
- University of Arizona Faculty newspaper, 25 November 1985: The lightning location system has earned more royalties for the university than all other university patents combined.
- Dr. Martin Uman, 1985: Elected Fellow of the American Meteorological Society.
- Dr. E. Philip Krider, 1985: The Award for Outstanding Contribution to the Advancement of Applied Meteorology given by the American Meteorological Society recognized his contribution for "the development and application of lightning detection instrumentation, thereby improving public services and helping to save lives and property."
- *Boston Globe* newspaper, 15 July 1985: Quotes a firefighter based at the National Interagency Fire Center in Boise that "I'm absolutely convinced that without these systems like lightning detection, we would have lost a lot more forest this year."
- Dr. E. Philip Krider, 1986: Elected Fellow of the American Meteorological Society.
- *Tucson Citizen* newspaper, 11 November 1986: Headline "Lightning zaps Arizona a record 217,298 times" is based on the BLM network. Note that the article has a lightning photo by Warren Faidley who is featured in Chap. 3.
- *EPRI Journal*, November 1986: The 26-sensor East Coast SUNYA network is proving to be valuable for the electric transmission and distribution system.
- *The Arizona Republic* newspaper, Phoenix, 17 August 1987: "LLP President Gary Arnold says that 'Most people don't know that you can detect lightning...'"
- *The Phoenix Gazette* newspaper, 30 September 1987: The article stresses that LLP is an example of basic research turned into a useful product. It quotes Mr. Lonnie Brown of the BLM in Idaho as LLP sensors are replacing a Learjet that flew a grid looking for lightning or fires in remote areas.
- *Science*, 27 October 1989: A landmark paper by Drs. Uman and Krider (1989) has a summary of the lightning phenomenon, impacts on rocket

launches, an NLDN map, and recommendations for future lightning research.

- August 1991: Numerous newspapers reported on LLP's competitor ARSI in the marketplace of lightning detection; ARSI was acquired by LLP in 1993.
- *The Arizona Daily Star* newspaper, 27 June 1991: GDS/LLP announced that it will build a data center for the NLDN in Tucson to be operational by June 1992.
- *St. Louis Post-Dispatch*, 12 August 1991: Published a national map of cloud-to-ground (CG) flash density from the NLDN for the year 1989.
- *The Arizona Daily Star* newspaper, Tucson, 12 April 1992: The Masters golf tournament in Georgia was stopped for the first time by data from lightning detection equipment. There had been two fatal golf incidents at tournaments in the previous year.
- *Electrical World*, February 1993: EPRI estimated at the time that lightning accounted for 33 to 50% of power failures in the U.S., and damages to power equipment was $50 million to $200 million a year. These expenses fully justified the cost of developing the NLDN.
- *Tucson Citizen* newspaper, 31 May 1993: Describes the NLDN operated by GAI as "Lightning Inc."
- *University of Arizona College of Science News*, Fall, 2000: Article titled "Application of College of Science Research Spawns a Global Corporation" shows continued pride in the development of real-time lightning detection.
- *The Arizona Daily Star* newspaper, Tucson, 21 August 2022: Article titled "Monsoon trackers get charge from new data set" describes how the National Weather Service offices in Arizona are using NLDN CG flashes to monitor state, county, and major city lightning through the monsoon season since there are remote regions with minimal coverage of other weather data.

This rapidly developing new technology and insights into lightning created the need for a conference specifically about lightning detection. Starting in 1979, meetings called ILDC took place in Tucson nearly every year. Since real-time lightning detection was a breakthrough, each year's conference was a benchmark when new results were presented about lightning detection technology and applications of network data. Basic and applied researchers who had not previously met were able to compare methods and applications. The conferences were science-based and did not directly involve marketing. Since networks based on the same technology as the NLDN were being established around the world, attendance from outside the U.S. soon reached

50%. Although there are other lightning conferences, ILDC was the only one dedicated to real-time lightning detection by members of the fields of physics, engineering, computation, meteorology, utilities, forestry, safety, aviation, geophysics, and protection. ILDC evolved into a 2-year cycle in 1996 in order to alternate with the American Meteorological Society Annual Meeting that has an embedded conference titled Meteorological Applications of Lightning Data. After Vaisala acquired GAI in 2002, ILDC moved around the U.S. outside of Tucson; one ILDC was held in 2002 in Vaisala's home base of Helsinki, Finland. One of the basic features of lightning research is that most people with interests in lightning have only one or a few people at a single location, or a person has lightning assigned as part of other duties. As a result, they are often not aware of applications of lightning data outside their immediate area of interest. ILDC provided a broader forum where lightning detection is at the core but the associated research and user communities resulting from detection can learn from each other in ways that may never have occurred to them. In 2006, the partner International Lightning Meteorology Conference (ILMC) relating to all types of applications was added to the original ILDC, so that ILDC/ILMC lasted 4 days in even-numbered years. Several thousand people attended ILDC/ILMC meetings.

6.4   Lightning Detection Businesses Based on Arizona Technology

As the technology was developed and improved, and business situations evolved, several companies were formed after the original invention of real-time lightning detection at the UArizona. Here is a brief overview of the companies and some highlights of the people involved:

- **Lightning Location and Protection Inc. (LLP)**: Once the demand arrived from the Alaska forestry interests, a company was founded in Tucson in 1976 by Drs. Uman and Krider. It readily became apparent that the contracts to build and maintain the sensors were not possible from a faculty office at the university. As a result, LLP was formed with an exclusive license from UArizona. Dynatech bought the company in 1983. Mr. Dick Lorimer was LLP President for 3 years, followed by Mr. Gary Arnold in 1987.
- **GeoMet Data Services (GDS)**: While LLP was focused on development, fabrication, and deployment of the sensors in the field, the demand for historical data was proving to be as important. As a result, the separate

business GDS was created in 1991 to provide data on a commercial basis to those examining insurance claims, utilities who were interested in evaluating lightning threats to their facilities, and many other interests who wanted to know when, where and how much lightning had occurred for a specific situation.

- **Global Atmospherics Inc. (GAI):** The combination of LLP, GDS, and ARSI into GAI was made in 1996 when Sankosha bought these separate companies. Another thread of lightning detection involved in the formation of GAI was the 1993 acquisition of time-of-arrival technology from the former ARSI. At that time, direction finding, and time-of-arrival were combined into the new IMPACT sensors.
- **Vaisala:** Based in Helsinki, Finland, Vaisala is a quality leader in nearly every type of weather observation used worldwide by meteorological agencies, airports, and many other interests. They recognized that GAI technology was the best in the world, so they acquired it in 2002.

As always in a new discipline such as real-time lightning detection, specific people make the difference. At the terrible risk of omitting obvious people, here are some major players in Tucson who strongly impacted the development of these companies:

- **Michael Maier:** He recognized the value of lightning data at the very beginning (Boulanger and Maier 1977; Maier et al. 1983; Maier and Jafferis 1985) from a user point of view and was employed by LLP for several years, including serving as Science Applications Manager starting in 1981. During his time with LLP, the number of direction finders in North America grew from none in 1975 to 170 in 1982 (Maier et al. 1983). He steered the first author of this book into realizing the potential of lightning data, and radically changed his career!
- **Leon Byerley III, Carl Noggle, and Dr. Burt Pifer:** They were pioneers in the development of the original detection technology and brought innovations from the latest technology to make the network actually work (Krider et al. 1976 Krider et al. 1980). Every high-tech company is built on a core of innovators such as Leon, Carl, and Burt.
- **Dr. Martin Murphy:** He started as a UArizona Ph.D. student of Dr. Krider, and transitioned full-time to LLP, GAI, and Vaisala in a variety of theoretical, technical, and meteorological roles that have been a cornerstone of the scientific reputation of the company (Murphy 1996; Cummins and Murphy 2009).

- **Patrick Zumbusch**: He arrived as president of LLP in 1989 and grew it quickly from its university-based facility into one that reached out in innovative ways for over a decade. The company grew from a dozen to 125 people during his tenure until purchased by Vaisala.
- **Dr. Ken Cummins**: He arrived at LLP in 1989 at about the same time as the IMPACT sensor was deployed. He served as Vice-President for Research and Development for many years. As described in the interviews in Sect. 7.4.2, he was recruited by Zumbusch in Wisconsin from a previous technology firm and continues lightning research at the UArizona (Cummins et al. 1992; Cummins et al. 1998).
- **Dr. Amitabh Nag**: He succeeded Dr. Cummins as Chief Scientist with Vaisala and continued the wide-ranging solid scientific collaboration to keep fully connected with the scientific community. He is now at the Los Alamos National Laboratory and co-authored the most complete summary of lightning detection that has been published (Nag et al. 2015).
- **Dr. Ryan Said**: He joined Vaisala from Stanford University, where he developed the hardware and software of what is now Vaisala's Global Lightning Dataset GLD360 network that covers the world as no other network has done in the past. He has rigorously continued the scientific basis of Drs. Krider, Cummins, and Nag (Said et al. 2010).

6.5   Sensor and Network Development and Deployment

The present-day sensors themselves containing both a direction finder and a GPS clock are not too impressive from the outside (Fig. 6.2). The critical aspect of the sensor is described in the patents and early papers. An antenna monitors the incoming electromagnetic emissions from all radio sources in the low-frequency (LF)/very low-frequency (VLF) band. When the signal satisfies five waveform criteria on the millisecond time scale, it is unambiguously identified as coming from lightning. One feature of the signal when lightning contacts the ground is specifically identified as the source of the unique time and angle from lightning (Cummins et al. 1998). One of the core principles is that when a CG flash comes in contact with the surface of the earth, the last step leader is almost exactly vertical, which simplifies the detection of angles (Uman et al. 1980). A processor within the antenna then organizes all that information into a message with the time to the microsecond, angle in degrees, signal strength converted to kiloamperes, and

Fig. 6.2 NLDN sensor near Yuma, Arizona. The GPS clock and direction-finding antenna are located within the small cap on top. The communications and power equipment are in a small hut positioned several hundred meters from the sensor (photo courtesy of Vaisala)

polarity, along with some features of the incoming waveform, and sends it by satellite to the central processor at the control center.

The first antennas with only a direction finder using the lightning waveform criteria were installed in Alaska (Fig. 6.3). Direction finding for radio navigation and other applications had a long history through the 20th century; among these were major studies for locating lightning by Stansfield (1947), Norinder (1954), and Horner (1957). An early test of direction finding of lightning over long distances by Lewis et al. (1960) took place prior to development of greatly improved electronics and communications. One of the first UArizona displays in Fig. 6.4 shows the workstation in Fairbanks, Alaska, along with a sample of an azimuth map (Noggle et al. 1976; Vance and Krider 1978). Figures 6.3 and 6.4 were provided from the personal library of Dr. Krider (see interview in Sect. 7.4.1.9).

In the interview with Dr. Krider (Sect. 7.4.1), it was recalled "And that was actually a useful tool for them, because they could vector the airplanes that go out and look for fires and they knew about how far away, and then they could tell the exact direction. Then they look for the fire." Two early direction-finding antennas were also deployed during 1978 and 1979 in the Tampa, Florida area for intercomparison with field mills, flash counters, cameras, and other instrumentation in order to test the quality of a lightning location system for utility applications (Darveniza and Uman 1982).

Fig. 6.3 The triangular white antenna is a very early direction finder installed at the former McGrath, Alaska BLM field station in 1976

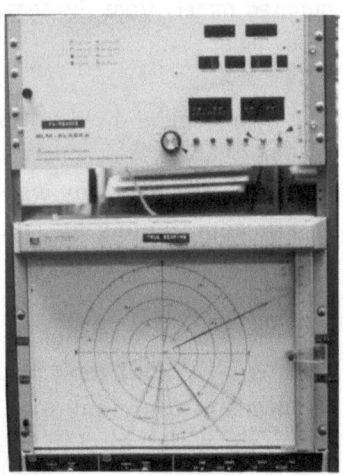
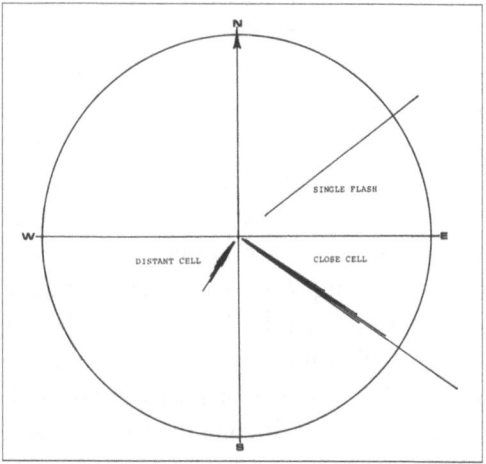

Fig. 6.4 Left: Workstation showing data from a single direction-finding antenna at the Fairbanks, Alaska BLM office station in 1976. Right: Azimuths and lightning types on August 2, 1976, from the Fairbanks antenna

How does a network of sensors such as that shown in Fig. 6.2 locate and timestamp a lightning event in the NLDN? Let's step through the process and describe briefly what is measured, where the sensors are located, and how a lightning event's location is calculated with these ideas:

- What is measured at each sensor? The time and angle from the antenna to the lightning event. At first, only angles were used, as in Figs. 6.3 and

6.4, then timing was added as GPS clocks became economically practical. A more complete description of the network and geometry involved in lightning detection is in (Cummins et al. 1998).

- Where are the sensors located? One of the best locations turns out to be places such as small airports. The sensors are extremely sensitive to the incoming signal in the LF/VLF frequency band. As a result, only about 110 antennas are needed to cover the U.S. with the present location accuracy of 100 meters or less (Nag et al. 2014; Zhu et al. 2020), and over 95% of CG flashes being detected. However, this sensitivity requires an electromagnetically quiet neighborhood. A small airport typically has a wide area without towers or other radio sources in the vicinity of the runways. Airports also have reliable power and reliable security to avoid physical damage such as from animals and theft. Any other site with these features can also serve as a site host. A buried cable connects the sensor to a small shed located some distance away where communications are maintained between the sensor and the operations center through a satellite dish.

- How is a lightning position found? The times, angles, and other information from each sensor arrive at a central processing center. Here, spherical trigonometry is used to locate the intersections of the sensor data within time and space constraints. The topography between the lightning occurrence and the sensor is also taken into account, since it takes longer for a lightning signal to travel over rough terrain than over flat land. After waiting a few seconds, the central Position Analyzer has enough information to calculate the lightning's location, signal strength, polarity, and measures how well it was located. Complex statistical optimization methods are used to find the best solution (Cummins et al. 1998).

Why Use Spherical Trigonometry?

Why is spherical trigonometry involved? Did you enjoy spherical trigonometry in high school or college? How much does the earth curve across the distance from lightning striking the ground to an antenna 250 miles away?

After the individual direction finders were developed and installed, a few local networks were placed in Florida (Darveniza and Uman 1979; Maier and Krider 1982; McGraw 1982; Byerley et al. 1983), the western U.S. (Western Region, NWS 1982a, b; Mielke 1990), Alaska (Hoblitt 1994), and the southern Plains (Goodman and MacGorman 1986) as well as outside North America such as in Sweden (Andersson and Jacobsson 1986) in the late 1970s to early 1980s. The success of these networks in operational applications as well as in lightning research was quickly realized, and the desire soon came to establish networks covering much larger areas.

The first large-network coordination center was established in the early 1980s at SUNYA (Orville interview in Sect. 7.4.3). Since lightning occurs at any time of day or night across the U.S., such a center needs to be staffed continuously onsite while a large number of real-time functions need to be monitored. These issues were resolved by Ron Henderson and others at SUNYA for the first time, using a linked bank of personal computers. Dr. Richard Orville, who had been at the UArizona for the early stages of this development, implemented the first real-time regional network control center (Orville 2008; Orville et al. 1986). EPRI funded the installation of this network so that a variety of power companies could benefit from tracking lightning over a larger area than a single utility (Maffetone et al. 1991; Bowden and Keener 1993; Bernstein et al. 1996), as well as aviation interests (Orville et al 1983a, b). The success of this east-coast EPRI network established by SUNYA set the stage for expansion into a national network. An unexpected discovery during the late 1980s was that some CGs lowered positive charge to ground (Brook et al. 1989; MacGorman and Taylor 1989). More context about this realization is in the interview with Dr. Orville in Chap. 7.

In 1992, the Network Control Center, later called the Network Operations Center (NOC) was established by GAI in Tucson. Figure 6.5 shows the NOC in late 2019 before it was re-located to the Vaisala office in Louisville, Colorado. For over 25 years in Tucson and now in Louisville, the operations center is staffed 24/7/365. What do they do? The sensors themselves rarely have outages—although a few have been struck directly by lightning over the years. However, the incoming power and the satellite-based communications need monitoring. Hardware and software upgrades occur, sometimes with unintended consequences, or a storm may cause local power to go down so that the sensor needs to be rebooted from the central facility.

A large number of papers have been published with results based on the NLDN for the entire U.S., such as benchmark entries by Orville and Silver (1997) and Cummins et al. (1998). The Canadian Lightning Detection Network (CLDN) was completed in 1998; while it is owned by Environment and Climate Change Canada, the CLDN has always been operated seamlessly with the NDLN by Vaisala from the NOC (Orville et al. 2002; Kochtubajda and Burrows 2020).

Fig. 6.5 Network Operations Center in Tucson in 2019. Several of the displays are related to lightning. (© R. Holle)

References

Andersson T, Jacobsson C (1986) The 3-dimensional distribution of radar reflectivity and lightning—a pilot study. Nordiska Meteorologmotet, 26–30 May, Visby, p 12

Bernstein R, Samm R, Cummins K et al (1996) Lightning detection network averts damage and speeds restoration. IEEE Comput Appl Power

Boulanger AG, Maier MW (1977) On the frequency of cloud-to-ground lightning from tropical cumulonimbus clouds. In: Proceedings of the 11th technical conference on hurricanes and tropical meteorology. American Meteorological Society, Miami Beach, Florida

Bowden G, Keener RN Jr (1993) Duke Power uses lightning network to reduce crew dispatch costs. EPRI Innovator IN-101090

Brook M, Henderson RW, Pyle RB (1989) Positive lightning strokes to ground. J Geophys Res 94:13,295–13,303

Byerley LG, Binford C, Maier MW (1983) The LLP lightning locating system. In: Preprints of the 9th conference on aerospace and aeronautical meteorology. American Meteorological Society, Omaha, Nebraska, 6–9 June 1982

Cummins KL, Murphy MJ (2009) An overview of lightning locating systems: History, techniques, and data uses, with an in-depth look at the U.S NLDN. IEEE Trans Electromagn Compat 51:499–518

Cummins KL, Murphy MJ, Bardo EA et al (1998) A combined TOA/MDF technology upgrade of the U.S. national lightning detection network. J Geophys Res 103:9035–9044

Cummins KL, Hiscox WL, Pifer AE et al (1992) Performance analysis of the U.S. national lightning detection network. In: Proceedings of the 9th international conference on atmospheric electricity, St. Petersburg, Russia

Darveniza M, Uman MA (1979) Lightning studies of transmission lines. In: Proceedings of the IEEE power engineering society conference on overhead and underground transmission and distribution, Atlanta, Georgia

Darveniza M, Uman MA (1982) Lightning protection of distribution lines; Final report. US Dept Energy, DOE/ET/29066-1, p 45

Goodman SJ, MacGorman DR (1986) Cloud-to-ground lightning activity in mesoscale convective complexes. Mon Wea Rev 114:2320–2328

Herman BD, Uman MA, Brantley RD et al (1976) Test of the principle of operation of a wide-band magnetic direction finder for lightning return strokes. J Appl Meteor 15:402–415

Hoblitt RP (1994) An experiment to detect and locate lightning associated with eruptions of Redoubt Volcano. J Volcanol Geothermal Res 62:499–517

Horner F (1957) Very-low-frequency propagation and direction-finding. Proc Inst Elect Eng 104, part B, 14:73–80

Idone VP, Orville RE (1990) Delimiting "thunderstorm watch" periods by real-time lightning location for a power utility company. Weather Forecast 5:139–147

Kochtubajda B, Burrows WR (2020) Cloud-to-ground lightning in Canada: 20 years of CLDN data. Atmos Ocean 58:316–332

Krider EP, Noggle RC (1974) Broadband antenna systems for lightning magnetic fields. J Appl Meteor 14:252–256

Krider EP, Noggle RC, Uman MA et al (1974) Lightning and the Apollo 17/Saturn V exhaust plume. J Spacecr Rocket 11:72–75

Krider EP, Noggle RC, Uman MA (1976) A gated, wideband magnetic direction finder for lightning return strokes. J Appl Meteor 15:301–306

Krider EP, Noggle RC, Pifer AE et al (1980) Lightning direction-finding systems for forest fire detection. Bull Amer Meteor Soc 61:980–986

Lewis EA, Harvey RB, Rasmussen JE (1960) Hyperbolic direction finding with sferics of transatlantic origin. J Geophys Res 65:1879–1905

Lewis J (1989) Real time lightning data and its application in forecasting convective activity. In: Preprints of the 12th conference on weather analysis and forecasting. American Meteorological Society, Monterey, California

Lin YT, Uman MA, Tiller JA et al (1979) Characterization of lightning return stroke electric and magnetic fields from simultaneous two-station measurements. J Geophys Res 84:6307–6314

MacGorman DR, Taylor WL (1989) Positive cloud-to-ground lightning detection by a direction-finder network. J Geophys Res 94:13,313-13,318

Maffetone T, Mark D, Montgomery W et al (1991) More accurate lightning data reduce utility's thunderstorm watch periods. EPRI Innovator IN-100026

Maier MW, Jafferis W (1985) Locating rocket triggered lightning using the LLP lightning locating system at the NASA Kennedy Space Center. In: Preprints of the 10th international conference on lightning and static electricity, National Interagency Coordination Group, Paris, France, 10-13 June 1985

Maier MW, Krider EP (1982) A comparative study of the cloud-to-ground lightning characteristics in Florida and Oklahoma thunderstorms. In: Preprints of the 12th conference on severe local storms. American Meteorological Society, San Antonio, Texas

Maier MW, Binford RC, Byerley LG et al (1983) Locating cloud-to-ground lightning with wideband magnetic direction finders. In: Preprints of the 5th symposium on meteorological observations and instrumentation. American Meteorological Society, Toronto, Ontario, Canada

McGraw MG (1982) 'On-line' lightning maps lead crews to 'trouble.' Elect World 196:111–114

Mielke KB (1990) Operational use of lightning data in western United States. In: Preprints of the 16th conference on severe local storms. American Meteorological Society, Kananaskis Provincial Park, Alberta, Canada

Murphy MJ (1996) The electrification of Florida thunderstorms. Phd dissertation, Department of Atmospheric Sciences, University of Arizona, p 134

Nag A, Murphy MJ, Cummins KL et al (2014) Recent evolution of the US national lightning detection network. In: Preprints of the 23rd international lightning detection conference, Tucson, Arizona

Nag A, Murphy MW, Schulz W et al (2015) Lightning locating systems: insights on characteristics and validation techniques. Earth Space Sci 2. https://doi.org/10.1002/2014EA000051

Noggle RC, Krider EP, Vance DL et al (1976) A lightning direction-finding system for forest fire detection. In: Proceedings, 4th conference on fire and forest meteorology. American Meteorological Society, St. Louis, Missouri

Norinder H (1954) The wave-forms of the electric field in atmospherics recorded simultaneously by two distant stations. Arkiv Geofys 2(9):161–194

Orville RE (2008) Development of the national lightning detection network. Bull Am Meteor Soc 89:180–190

Orville RE, Silver AC (1997) Lightning ground flash density in the contiguous United States: 1992–1995. Mon Wea Rev 125:631–638

Orville RE, Henderson RW, Bosart LF (1983a) An east coast lightning detection network. Bull Am Meteor Soc 64:1029–1037

Orville RE, Huffines GR, Burrows WR et al (2002) The North American lightning detection network (NALDN)–first results: 1998–2000. Mon Wea Rev 130:2098–2109

Orville RE, Henderson RW, Pyle RB et al (1983b) The use of the east coast lightning detection network in operational aviation meteorology. In: Preprints of the 2nd conference on the aviation weather system. American Meteorological Society, Montreal, Quebec, Canada

Orville RE, Pyle RB, Henderson RW (1986) The east coast lightning detection network. IEEE Trans Power Syst PWRS-1:243–246

Orville RE Jr (1987) An analytical solution to obtain the optimum source location using multiple direction finders on a spherical surface. J Geophys Res 92:10,877–10,886

Western Region, National Weather Service (1982a) Experimental lightning detection charts help locate thunderstorms. Tech Attach 82-32, 27 July, p 5

Western Region, National Weather Service (1982b) The ARAP, ALDS, and GOES meet in southern Utah. Tech Attach 82-34, 3 August, p 4

Said R, Inan U, Cummins KL (2010) Long-range lightning geolocation using a VLF radio atmospheric waveform bank. J Geophys Res 115:D23108

Salanave LE (1961) The optical spectrum of lightning. Science 134(3488):1395–1399

Salanave LE (1980) Lightning and its spectrum: an atlas of photographs. University of Arizona Press, Tucson, p 136

Slipher VM (1917) The spectrum of lightning. Lowell Obs Bull 79, 3(4):55–58

Stansfield RB (1947) Statistical theory of D.F. fixing. J IEEE, Part IIIa 94:762–770

Uman MA, Krider EP (1989) Natural and artificially initiated lightning. Science 246:457–464

Uman MA, Lin YT, Krider EP (1980) Errors in magnetic direction finding due to nonvertical lightning channels. Radio Sci 15:35–39

Vance DL, Krider EP (1978) Lightning detection systems for fire management. In: Preprints of the 5th joint conference on fire and forest meteorology. American Meteorological Society and Society of American Foresters, Atlantic City, New Jersey

Walker TD, Christian HJ (2014) Novel observations in lightning spectroscopy. In: Preprints, XV international conference on atmospheric electricity, Norman, Oklahoma

Whitehead J, Driggans R (1990) TVA's experience with the SUNYA lightning detection network. IEEE Trans Power Deliv 5:6

Whitney BF, Asgeirsson H (1991) Lightning location and storm severity display system. IEEE Trans Power Deliv 6:1715–1720

Zhu Y, Lyu W, Cramer J et al (2020) Analysis of location errors of the US national lightning detection network using lightning strikes to towers. J Geophys Res 125(9):e2020JD032530

7

Lightning Research in Arizona

Abstract Tucson and Arizona have had significant roles as world-class lightning research and operations centers of excellence for several decades, and it is the home of the basic scientific and technological advances that led to modern real-time lightning detection from ground-based sensors (Chap. 6). The University of Arizona (UArizona) has been a trailblazer in many aspects of lightning research during studies in Arizona and elsewhere. A remarkable number of lightning scientists that are globally and nationally recognized have been employed or studied at UArizona, and their contributions are listed in this chapter. The chapter concludes with excerpts from in-person interviews by the authors with Drs. E. Philip Krider, Kenneth L. Cummins, and Richard E. Orville, three of the leading scientists involved in the invention, startup, deployment, and improvements that became the National Lightning Detection Network (NLDN). An extensive list of references related to these developments is also provided.

7.1 University of Arizona Lightning Studies Before 2000

A wide variety of basic and applied lightning studies were made at the University of Arizona (UA, now UArizona) before 2000, as shown throughout this chapter. In this chapter, we emphasize three research areas that deserve special mention.

© Springer Nature Switzerland AG 2023
R. L. Holle and D. Zhang, *Flashes of Brilliance*,
https://doi.org/10.1007/978-3-031-19879-3_7

7.1.1 Spectroscopy

Lightning studies in Arizona, especially at UArizona, can be traced back to research conducted in the early 1960s. Relevant to lightning was a focus at that time in the area of instrumentation, as the Department of Atmospheric Physics had an emphasis on all types of weather instruments, such as an anemometer by Kassander and Stewart (1955). One of the specific topics was spectroscopy to measure the temperature of lightning that was not well known at that time (Prueitt 1963; Uman 1964; Krider 1973; Weidman et al., 1989). Temperature affects the amount of lightning damages, and protection against lightning damages means that one must know the underlying cause of the damage. Spectroscopy is described in detail in the book by Salanave (1980) that has color photos of lightning and its spectrum. While Dr. Salanave was conducting these studies, he collected relevant papers into two blue binders called "Die Blitz-Bibel", the lightning Bible, as shown in Fig. 7.1.

Why Die Blitz-Bibel?

Remember that at that time, hard copies of reprints were collected by mail, by contacting colleagues by phone, and subscribing to journals, so this was not an easy task in the emerging but dispersed field of lightning studies.

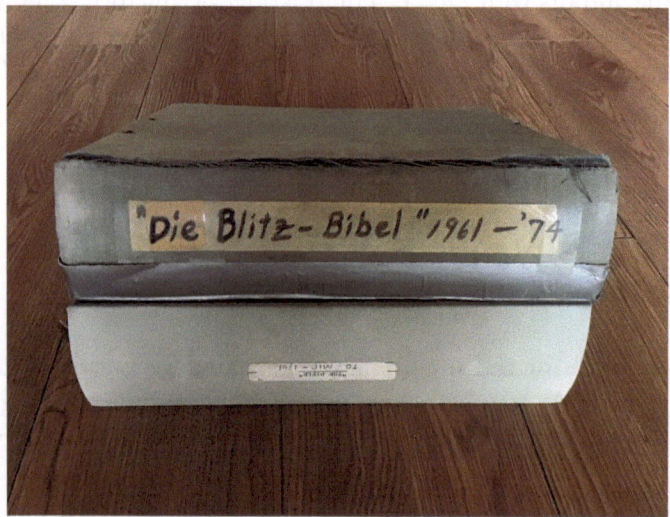

Fig. 7.1 The two volumes of "Die Blitz-Bibel" compiled by Dr. Leon Salanave that contain early publications related primarily to lightning spectroscopy (© D. Zhang)

7.1.2 Kennedy Space Center Lightning Studies

On November 14, 1969, the Apollo 12 space rocket was struck by lightning twice; one event was a cloud-to-ground (CG) flash, and the other was an in-cloud (IC) flash. This lightning incident caused a major electrical disturbance that temporarily disconnected fuel cells and left nine temperature/pressure sensors permanently damaged; the astronauts reported seeing and feeling the strike (Godfrey et al. 1970; Uman and Krider 1989; Krider interview in Sect. 7.4.1). No lightning stroke was measured or observed, and no thunder was heard during the six-hour periods before and after the launch except for the one CG that struck the rocket and an IC about 15.5 s afterward. Investigators concluded that the flash was most likely initiated by the rocket itself when it was flying through an electrified cloud. The initial inquiry led to further lightning-related measurements, observations, and experiments, such as the Thunderstorm Research International Project (TRIP, Pierce 1976), as well as drastically raising lightning safety awareness for aviation and rocket launches. The investigators included:

- Dr. Louis J. Battan from the UArizona Institute of Atmospheric Physics (UArizona IAP),
- Dr. E. Philip Krider while he was a NASA-NRC (National Research Council) Resident Research Associate at the Manned Spacecraft Center,
- Dr. Martin A. Uman from the Research and Development Center at the Westinghouse Electric Corporation, who worked as a young faculty member in the Department of Engineering at UArizona during 1961–1965, and
- Dr. Richard E. Orville from the Department of Atmospheric Science at the State University of New York at Albany, who was a graduate student in the Department of Atmospheric Sciences at UArizona during 1961–1966.

It was during the aftermath of this landmark event that seeds of lightning detection were planted. This development, described in Chap. 6, took the existing basic research of lightning, and converted this knowledge into the operational lightning detection methods that are so common today (Oram et al. 2005).

For a decade, the UArizona lightning team drove to the Kennedy Space Center (KSC) every summer for field campaigns with a UArizona IAP school bus, known as the "lightning bus." This lightning bus was about half the length of a normal school bus (Fig. 7.2) and was an old Navy vehicle originally painted dark blue. The UArizona lightning team painted it white to try to keep the inside cool in the Florida summer heat. The inside of the light-ning bus was very congested (Fig. 7.3). There were no waveform digitizers back then, but a big oscilloscope with cameras to photograph waveforms as they swept across the oscilloscope screen (Weidman and Krider 1986). Dr. Charles Weidman, who was involved in the field campaign mentioned:

> Mostly we filled the bus with equipment and then drove it to Florida every summer. We would park it at one of the field mill sites near the Apollo launch pads (39A and 39B). It would sit there for a couple of months and then we'd drive it back to Tucson. It was a long drive each way, about five days. The bus didn't have any air conditioning and would break down at some point every year. But it always made it back to Tucson.
>
> One summer rather than going to the Kennedy Space Center we parked the bus on the west coast of Florida near St. Petersburg. Another year after the summer field experiments were over in Florida we drove the bus to Roberval, Quebec, Canada, and did two weeks of whistler experiments in collaboration with a research group from Stanford. We actually lived in the bus during that time, since all the hotel rooms in the nearby town were booked up.
>
> After the summer field experiments, we'd leave the bus at the UArizona farm. One year somebody kind of moved into the bus and turned it into their own personal "club house." This person kept the bus clean. Eventually someone else found out about the bus and vandalized it pretty badly. That was the end of the lightning bus.

KSC maintains the largest real-time electric field mill network in the world. It is expensive but worth the investment in hardware, software, main-tenance, and especially the development of understanding since field mills are a critical part of the lightning detection and warning system at KSC. Data from the network have been examined for decades by faculty and staff at UArizona, particularly Dr. Krider. One of the fundamental studies with the KSC field mill network was made for a Ph.D. dissertation by Dr. Martin Murphy, now at Vaisala Inc. in Colorado. He described the capabilities and limitations of a field mill network and analyzed frequency, duration, thresholds, and magnitudes of field mill lightning events at KSC (Murphy 1996). To this day, the understanding gained from this, and other UArizona research such as Maier and Krider (1986) forms an integral part of lightning nowcasting methods at KSC. The Launch Commit Criteria are extremely detailed and most of the operational attention is focused on lightning (Merceret et al. 2010).

Fig. 7.2 The UArizona Institute of Atmospheric Physics lightning team and the light-ning bus located at the northern end of KSC. From left to right: Joel Hamelin, Dr. Martin Uman, Dr. Philip Krider, Dr. Charles Weidman, Dr. Changming Guo, and B. Djebari (© Dr. Charles Weidman, UArizona)

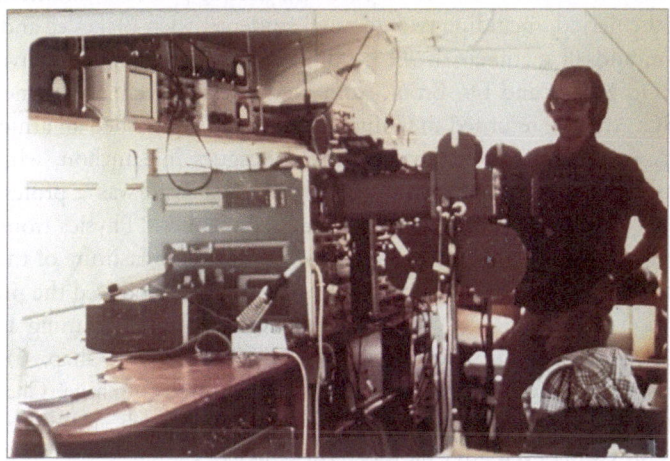

Fig. 7.3 Dr. Weidman inside the lightning bus (© Dr. Charles Weidman, UArizona)

Rocket Launch Delays

You probably have heard about lightning canceling or delaying rocket launches at KSC and other space facilities around the world. The basic and applied knowledge about lightning is an integral part of making sure that no rocket initiates lightning while being launched into seemingly innocuous clouds.

One of the many prominent visitors to UArizona during this time was Dr. Changming Guo. The following summary of Dr. Guo's involvement at UArizona and later in China is described by Dr. Xiushu Qie, an early student of Dr. Guo, and currently a leader of lightning research at the Institute of Atmospheric Physics of the Chinese Academy of Sciences in China:

Changming worked in Dr. Krider's group at UArizona from January 1980 to October 1982 as a visiting scholar from Lanzhou Institute of Plateau Atmospheric Physics, Chinese Academy of Sciences. His research interests during his visit were lightning physics and detection. Supervised by Dr. Krider, he assembled a set of lightning optical imaging sensors with fast time resolution by taking advantages of the university's laboratory facilities. He was appreciative of the opportunity to take part in the summer lightning campaign at KSC as a member of the group. Based on the data collected about lightning's optical and radiation fields, three papers were completed and published in the *Journal of Geophysical Research*. The optical power radiated by return strokes and the energy dissipated by lightning were estimated, and the anomalous light output from dart leaders was discovered. During the stay in the group, he also sat in Dr. Krider's classes. Many colleagues and friends provided additional assistance in the period, including two doctoral students, Rich Blakeslee and Chuck Weidman, and all members in the group, Prof. Uman from the University of Florida, Dr. Moore and Dr. Brook from New Mexico Tech, and some friends from KSC. After he returned to China, Changming established an atmospheric electricity research group with his young colleagues in Lanzhou, which later became well known in the world. Dr. Changming Guo was a professor and director of the Lanzhou Institute of Plateau Atmospheric Physics from 1986–1993 and became the director of the Shanghai Typhoon Institute of the China Meteorological Administration from 1994–1998. He introduced the prototype three-station lightning location system into China from Lightning Location and Protection (LLP) in the middle 1980s (described in Chap. 6), which significantly promoted lightning detection and research in China. Changming has been continuously dedicated in advancing the research and protection of lightning in China even after he was retired in 1999.

7.1.3   Lightning on Other Planets

Lightning does not only occur on Earth but also on other planets, such as Jupiter (Williams 1986). However, whether there is lightning on Venus is still controversial, so additional studies have attempted to answer the question. Researchers from the UArizona Lunar and Planetary Laboratory used the 153-cm telescope on Mt. Bigelow in the Santa Catalina Mountains on the north side of Tucson to show supporting evidence (Hansell et al. 1995;

Lorenz 2018). This was not only one of the earliest attempts for finding lightning discharges on Venus but also was designed to look systematically for lightning there.

7.2 University of Arizona Lightning Studies After 2000

7.2.1 Lightning-Rainfall Studies

As the NLDN was becoming operational over the U.S., many applications were tested. One of the more prominent questions was the ability of lightning to estimate rainfall. There are various situations where radar, rain gage, and satellite data are not completely reliable in estimating rainfall, so knowing the practicality of a lightning-rainfall relationship is very useful. A combination of meteorological sensors is best when and where possible. However, in many areas of the world, there are no radar data, and satellite data are not easily available with the timeliness and spatial resolution to make accurate estimates of rainfall for floods, agricultural applications, drought monitoring, and other situations. For that reason, Kempf and Krider (2003) examined rainfall over the central U.S. in a major flood event. The relationship was very good, but lightning data by themselves have limitations since some storms can produce a large amount of rain with little lightning, and vice versa.

Over Mexico, there is a significant lack of real-time radar and other data accessible to meteorological and hydrological organizations. The use of lightning data to assist in the estimation of rainfall was pursued for a Ph.D. by Carlos Minjarez-Sosa from the University of Sonora in the Mexican state directly south of Arizona. A series of studies by Minjarez-Sosa et al. (2012, 2017, 2019) and coauthors from UArizona and the National Autonomous University of Mexico (UNAM) in Mexico City explored the lightning-rainfall relationship for quantitative precipitation estimation with the NLDN. The aim was to extend the results of Kempf and Krider (2003) in the central U.S. plains to the complex terrain of the southwest U.S. and northwest Mexico. These studies compared lightning data with both rain gage and radar data.

Researchers in other countries continue to build on the results of Kempf and Krider (2003) and Minjarez-Sosa et al. (2012, 2017, 2019) to explore the factors in specific regions with varying meteorological regimes. A vexing problem is that not all rainfall is accompanied by frequent lightning, such as during the large mesoscale convective systems described in Sect. 1.5.3. These monsoon-season complexes have stratiform rain where the rain rate is lower, but the duration is long, and the area is large. In such regions with a low rate of widespread lightning, a different lightning-rainfall relationship needs to be used but doing so objectively is problematic. Nevertheless, there are promising practical applications in developing countries where radar may not become available, and satellite data are not likely to be fully accessible in the foreseeable future.

7.2.2   Grand Canyon Lightning Studies

Arizona is the home of the world-famous Grand Canyon. Its unique landscape makes it a treasure of the geological history of tens of millions of years and attracts visitors from all over the world. During the summer monsoon season, thunderstorms develop in northern Arizona and occur within Grand Canyon National Park (Holle et al. 2021). Lightning frequency in the Grand Canyon area was addressed in Sect. 4.4.1, and casualties in Sect. 5.1.7. Now we focus on the unique factors involved in lightning formation in and around this giant chasm.

In 2009, UArizona scientists carried out an observation campaign at the Grand Canyon to study how lightning interacts with complex terrain (Fig. 7.4). The field campaign involves utilizing lightning videos taken from the Grand Canyon by then a Ph.D. student Mason Quick, along with the NLDN, radar, and other meteorological data to explore the effects of complex terrain over lightning. Typically, thunderstorms form on the elevated terrain along the south rim of the canyon and move northward. However, Dr. Cummins and his M.S. student Tyler Kranz found that most of these storms weaken or dissipate when crossing the canyon floor. This thunderstorm weakening may cause the abrupt drop in the number of CGs on the canyon floor (Figs. 4.22 and 4.23). Rarely, thunderstorms cross the canyon without weakening or dissipating, and such storms are either nocturnal or associated with large vertical wind shear over the canyon, which favors storm development and maintenance. No storm was found to initiate or show strengthening over the canyon floor. Therefore, lightning is most likely to strike along the rims but not the canyon floor.

Questions raised relate to (1) a deficit of lightning inside the depths of the canyon and (2) more lightning on the promontories than on the plateau. Several ideas proposed by Cummins in an unfunded proposal subsequent to 2009 and the M.S. study by Kranz are:

(1) "There is a clear increased incidence of lightning attachment near the top edge of the canyon rim. If the attachment is made right at/near the top

Fig. 7.4 CG lightning inside the Grand Canyon as captured by an automated camera during the 2009 UArizona field experiment. The red ball on the left is part of the clock system. Courtesy of Dr. Mason Quick

of the rim, this could imply a very large attractive radius produced by unusually long upward leaders, resulting from enhanced electric fields at the canyon rim just prior to ground termination."

(2) "If the attachment is made along the face of the canyon near the rim, which is the other possible interpretation of existing data, then the high stroke density near the rim and the low density within the canyon might be explained by the spatial distribution of downward branches just prior to attachment, interacting with the slope and protrusions along the canyon walls. This could cause attachment to the canyon walls to occur before lower-altitude leader channels made their way to the canyon floor."

(3) Do thunderstorms weaken as they cross the canyon? Most often, storms move across the canyon from south to north during the monsoon. Kranz and Cummins (personal communication) found that radar echoes crossing the canyon had difficulty maintaining their normal life cycle. As they crossed the canyon that is many km (miles) wide, a cloud may spend five or ten minutes over an area where the lower surface that was important for propagation by a gust front and surface heating is suddenly missing. In fact, most storms leaving the South Rim studied by Kranz and Cummins did not regenerate on the North Rim.

(4) If lightning strikes inside the canyon, is a lightning detection system missing them because they are not normal CG flashes?

Safety

Although it's not clear why lightning is reduced within the canyon, lightning occurs very frequently on the rim, so lightning awareness is critical, especially on promontories and rim locations where people congregate in large numbers.

7.2.3   Ground Truth Calibration of the NLDN

The development of the NLDN from a research topic to a business is described extensively in Chap. 6. Nevertheless, part of the success of the NLDN is due to ground truth about where lightning occurred, its strength in kiloamperes, polarity (negative or positive), and whether it was CG or IC. Many of these parameters had not been measured for a long period over a wide area with prior instrumentation. As a result, a series of studies was undertaken by UArizona faculty and staff, nearly always in combination with other researchers around the U.S. Ground truth took the form of camera studies (Biagi et al. 2007; Fleenor et al. 2009; Cummins et al. 2014a, 2014b), object strikes (e.g., towers, wind turbines, Warner et al. 2012; Cramer and Cummins 2014; Cummins et al. 2014b; Candela Garolera et al. 2015; Zhang et al. 2015), and complementary high-precision lightning sensors (Zhu et al. 2016; Zhang et al. 2016). They were often conducted with limited funding in experiments similar to the KSC bus project described in Sect. 7.1.2.

Artificial triggering of lightning is an experiment that has been conducted for over five decades in the U.S. (Uman and Krider 1989). Other countries such as France, Japan, Brazil, and China have also conducted similar experiments. Typically, a small rocket with a thin wire attached to it is launched during a thunderstorm. A team of researchers monitors the electric field to determine the best time to launch the rocket. The wire is disintegrated by the lightning event while the rocket falls back to the ground. The triggered lightning does not have a first stroke in the same way as the first return stroke of a natural flash (Sect. 1.4.2), but it is a very useful way to improve understanding of lightning from both a theoretical and practical point of view. The University of Florida conducted rocket-triggered lightning launches for many years in north Florida. Dr. Martin Uman, one of the founders of LLP in Tucson (Chap. 6) was a leader of this facility for a long time, and UArizona faculty and students took part in analyzing these data for research and NLDN applications (Uman and Krider 1989). Given the exact location, triggered lightning is often used for NLDN evaluation studies (Jerauld et al. 2005; Nag et al. 2011; Mallick et al. 2014).

7.2.4   Optical Studies of Lightning

Lightning return strokes release a large amount of power and radiation. Quick and Krider measured and examined the optical radiation emitted by return strokes in both natural lightning (Quick and Krider 2013) and rocket-triggered lightning (Quick and Krider 2015). A typical negative return stroke (the most frequent type of return stroke) produces optical signals with a fast rise-to-peak and slower decline feature. Sub-processes such as branching and M-components in the negative CG flashes can produce more complicated variations in the channel luminosity (Mach et al. 2005). Atmospheric extinction effects including scattering and absorption by aerosols can also have an impact on the measured optical energy (Orville and Henderson 1984). In general, both negative first return strokes and subsequent strokes with a new ground contact release a total optical energy of about 3.5×10^6 J, whereas positive return strokes release only one third as much energy. Rocket-triggered lightning, which reproduces almost the exact behaviors and processes as natural lightning (although the "first stroke" of a triggered lightning is different from the one of a natural lightning), has comparable power and energy to some natural lightning. The optical power radiated by triggered lightning is in good agreement with similar measurements made on the subsequent return strokes in natural lightning that remains in a pre-existing channel (Quick and Krider 2015).

In-Depth Studies

Projects such as the rocket-triggered program show how graduate students typically spend long hours collecting data in the field, still more hours preparing the data for analyses, and even longer hours analyzing the results.

7.2.5   Kansas Windfarm Project

Lightning is a major threat in the wind energy industry, since lightning can cause huge damage to wind turbine blades. If struck, a wind turbine blade sometimes needs to be repaired or replaced, and the average repair and replacement costs can be tens to hundreds of thousands of dollars. To minimize the potential risks, researchers at UArizona, along with scientists and engineers from other institutions and organizations, carried out a 3-month observational study at a windfarm in central Kansas to understand lightning attachment to wind turbines (Cummins et al. 2014a). They found that a

Fig. 7.5 Lightning striking a wind turbine during the UArizona field experiment in Kansas

direct strike from a typical negative CG stroke with only a moderate peak current (~30 kA) can cause severe damage (Wilson et al. 2013). Videos were also used (Fig. 7.5) to validate the update of the NLDN network that was made during the spring and summer of 2013 (Cummins et al. 2014b).

7.2.6   Lightning Studies from Space

The recent decades saw the rapid development of space-based lightning observations. Since the success of the Nighttime/Daytime Optical Survey of Lightning (NOSL) mission (Vonnegut et al. 1980), observing lightning from space (above the thunderstorms) became possible. Several satellite missions have then carried lightning sensors on either low Earth orbit (LEO) or geostationary orbit (GEO). These lightning sensors are mostly optical imagers that detect the optical signals from the transient pulses/strokes of total lightning (both CG and IC lightning). Unlike ground-based measurements, the satellite-based imagers capture the cloud-top lightning optical illumination that is scattered out from the cloud body and provide an additional insight into our understanding of lightning activity globally and regionally (Goodman et al. 2013; Cecil et al. 2015; Blakeslee et al. 2020).

By the time of the publication of this book, there will be five operational lightning imagers in space:

- Geostationary Lightning Mapper (GLM) on the GOES-16 satellite (launched in 2016),
- GLM on GOES-17 satellite (launched in 2018),
- GLM on GOES-18 satellite (launched in 2022),
- Lightning Imaging Sensor (LIS) on the International Space Station (ISS, placed in 2017),
- Lightning Mapping Imager (LMI) on the FengYun-4A satellite (launched in 2016).

Another lightning imager in preparation is the Lightning Imager (LI) on the future Meteosat Third Generation (MTG)-Imager satellite. Note that GLMs, LMI, and the future LI are/will be in GEO orbits whereas ISS-LIS is in a LEO orbit.

UArizona scientists have greatly contributed to space-based lightning studies. One of the primary works is to facilitate NOAA and NASA in satellite instrument calibration, validation, and evaluation studies. Zhang et al. (2019) evaluated the performance of the LIS on the Tropical Rainfall Measuring Mission (TRMM) satellite, which is the precursor of the other satellite sensors, such as GLMs. The study solved the problem of the location offset of the TRMM-LIS and developed the correction method for the data products after 17 years of satellite operation. Other instrumental issues of TRMM-LIS and ISS-LIS and how these issues can affect the instrument's detection, as well as how to make improvement were also studied (Zhang et al. 2019, 2021; Zhang and Cummins 2020; Zhang 2020). These works are significant to the international lightning community, as LIS data have been widely used for studying global lightning activity and lightning climatology. In addition, the key to creating a cohesive global satellite-based lightning dataset is to have reliable and consistent LEO sensors to provide validation, cross-calibration, and detection enhancement. Although GEO sensors developed by different countries share most of the sensor specifications, there are still large differences in the data products provided by each individual sensor, due to varying sensor resolutions and viewing geometries. Both LIS sensors are now under study in preparation for intercomparing and fusing different GEO and LEO satellite lightning sensors (Zhang et al. 2021).

Another interesting study by Medici et al. (2017) investigated a parameter called the IC lightning fraction in the Contiguous United States (CONUS) from four years of TRMM-LIS data. This parameter is useful for studying the physics of storm climatologies, as some storms tend to produce more IC flashes, whereas others make more CGs, largely due to the electrification structure of the storms. It is also useful for simulating the production of nitrogen oxides (NOx) by lightning called LNOx, since lightning is one of the largest natural NOx sources as in Jourdain and Hauglustaine (2001), and Gressent et al. (2014), which is also discussed in Sect. 5.4. An additional LIS application is that given the global coverage that is only within ±54° latitudes, ISS-LIS data were used to study the global lightning optical properties and land/sea differences (Cummins 2021). It was found that much larger and optically "brighter" flashes are mostly over ocean regions, compared with the land areas.

As a GEO sensor, GLMs are the first to detect the evolution of lightning activity from space throughout the lifecycle of thunderstorms that enables the observations of megaflashes (Peterson and Liu 2013; Lang et al., 2017; Peterson et al., 2022). Megaflashes are the largest flashes that have been observed, which can propagate hundreds of kilometers/miles horizontally (Sect. 7.3.2). In addition, the GLM data products provide a lead time for improving severe weather nowcasting such as tornadoes, strong winds, and large hail (Schultz et al. 2009, 2011). It has been noticed that the GLMs have a reduced detection efficiency during vigorously convective and/or inverted-polarity storms. Zhang and Cummins (2020) explained that these storms tend to have high flash rates and produce small-size flashes, and/or optically dim pulses due to thick clouds. These small and dim flashes underfill a GLM pixel footprint, which might cause decreased GLM detection.

In addition, UArizona researchers and engineers have helped optimize the GLM sensors. Both GLMs on the GOES-16 and -17 satellites were back-side processed at the University of Arizona Imaging Technology Laboratory (https://www.itl.arizona.edu/). This process optimizes the backside illumination of the GLM charge-coupled device (CCD) and improved their quantum efficiency by acid etching the surface of the sensors.

7.2.7 Fulgurites

Fulgurites are naturally produced glassy materials. The origin of the word "fulgurite" is the Latin word "fulgur," meaning lightning. When a CG flash strikes ground, the high temperature sometimes can melt sands, rocks, or other sediments, leaving elongated and round, sometimes hollow structures (Fig. 7.6). Also see the discussion of fulgurites in Sect. 5.3.10. The unique shape of a fulgurite is related to how the lightning current flows into the material and other less well-known factors. The most common elongated structure is mainly caused by the lightning current vaporizing the inner core material. A master's student in the Department of Planetary Science at the University of Arizona, Kristin M. Block, examined a collection of fulgurite morphology and developed the first classification system for fulgurites that is now widely used (Block 2011).

Fig. 7.6 Fulgurites obtained at the Tucson Gem and Mineral Show that were collected in Egypt. These examples are about 3 cm long (© R. Holle)

7.2.8 University of Arizona Lightning Studies by Students

As part of the KSC, Grand Canyon, NLDN, rainfall, and windfarm studies of lightning, a long series of Master's and Doctoral degrees were awarded by the University of Arizona (Table 7.1). These students not only furthered the understanding of the basic features of lightning, but many of the results had significant near-term applications. Table 7.1 includes primary publications related to university studies; this is not meant to be a complete representation of the impacts of each student's research, but hopefully, the list includes the main topics.

In 2016, the historical Department of Atmospheric Sciences started merging with the Department of Hydrology, and eventually became the Department of Hydrology & Atmospheric Sciences. Lightning and atmospheric electricity studies were no longer a departmental priority and the legacy of lightning studies ceased. Kranz and Zhang were the only two lightning students who graduated with the new department. No new lightning studies involving students are being initiated within the department at this time.

Table 7.1 University of Arizona lightning students arranged by year of key related publications

Student	Degree	Topic	Key early publication
Elizabeth Jacobson	M.S	KSC field mills	Jacobson and Krider (1976)
Charles Weidman	Ph.D.	Detailed lightning structure	Weidman (1982)
Michelle Piepgrass	M.S	KSC field mills/rain	Piepgrass and Krider (1982)
Launa Maier	Ph.D.	KSC LDAR	Maier et al. (1984, 1995)
Richard Blakeslee	Ph.D.	Maxwell currents	Krider and Blakeslee (1985)
Mark Williams	Ph.D.	Jupiter lightning	Williams (1986)
Thomas Adang	Ph.D.	Monsoon lightning	Adang (1989)
William Koshak	Ph.D.	KSC field mills	Koshak and Krider (1989)
Martin Murphy	Ph.D.	KSC field mills	Murphy (1996)
William Valine	M.S	Camera studies	Valine and Krider (2002)
Nicole Kempf	M.S	Rainfall versus lightning	Kempf and Krider (2003)
Natalie Murray	M.S	Stroke phenomena	Murray et al. (2005)
Bruce Gungle	M.S	Rainfall versus lightning	Gungle and Krider (2006)
Chris Biagi	M.S	NLDN validation	Biagi et al. (2007)
Kenneth Kehoe	M.S	NLDN validation	Biagi et al. (2007)
William Scheftic	M.S	Soil moisture and lightning	Scheftic et al. (2008)
Stacy Fleenor	M.S	Camera studies	Fleenor et al. (2009)
Lesley Leary Mozzarella	Ph.D.	Tropical cyclones	Leary and Ritchie (2009)
Christina Stall	M.S	Camera studies	Stall et al. (2009)
Mason Quick	Ph.D.	Optical studies	Quick and Krider (2013)
Gina Medici	M.S	Cloud lightning detection	Medici et al. (2017)
Carlos Minjarez-Sosa	Ph.D.	Rainfall versus lightning	Minjarez-Sosa et al. (2017)
Tyler Kranz	M.S	Grand Canyon	Presentations only
Daile Zhang	Ph.D.	Satellite lightning	Zhang et al. (2019, 2020)

Other international scientists who visited the lightning program at UArizona include the following, in addition to Dr. Changming Guo mentioned in Sect. 7.1.2:

- Christiane Leteinturier was a visiting scientist from France for a year to measure fast lightning E field changes at Cape Canaveral (Krider et al. 1992; Leteinturier et al. 1990).
- Leandro Zanella S. Campos was a visiting student from Brazil. He spent several months in 2007 for field work and in 2014 for visiting scholar time as part of his Ph.D. study. The topic of his doctoral research was multiple-ground terminations and clustering algorithms for lightning locating systems (LLS) data (Campos 2016).

7.3 Other Arizona Lightning Studies

7.3.1 Flagstaff Study

Most Arizona lightning activities in recent decades have been conducted either at UArizona or connected with the development of the NLDN by LLP and Vaisala; both took place in Tucson. A recent biography of the famous British cloud physicist Dr. John Latham mentioned that in 1966, he collaborated in aircraft studies of Flagstaff thunderstorms. Electric fields and particle charges were measured, and it was concluded that the electrical characteristics of complex clouds are often explicable in terms of the Reynolds-Brook theory (Reynolds et al. 1957). Nothing further has been found about this study; however, it reinforces how Arizona cumulonimbus clouds growing over isolated mountainous terrain can be a classic form of thunderstorms to develop foundational knowledge for other locations with more complex storm situations.

7.3.2 Arizona State University

On February 1, 2022, the World Meteorological Organization (WMO) announced the new records of lightning extremes, and these results have been published (Peterson et al. 2022). The longest horizontal distance a lightning flash traveled was reported as 768 km (477 mi) across the southern U.S. Gulf Coast on 29 April 2020. The longest duration a flash lasted was 17.01 s over the Argentina-Uruguay border on 18 June 2020. Thanks to the new GLM satellite, these records greatly extend the previous records published in 2017 (Lang et al. 2017), which were 321 km (199.5 mi) and 7.74 s, respectively. Dr. Randall Cerveny, a professor at Arizona State University and the chief Rapporteur of Weather and Climate Extremes for the WMO, was involved in both studies. This work has a tremendous impact on how we should reexamine our current lightning safety rules. One of them is the 30–30 rule that has two parts. The first part is if the time between flash and thunder is less than 30 s (which means that the person is less than 10 km from the flash origin), go inside! The second part is to wait 30 min after the last observed flash to resume outdoor activities. These are also described in the lightning safety Sect. 5.2. Although most lightning flashes will not travel hundreds of miles, those "megaflashes" do exist and they can travel several hundred miles within a few seconds. This can be a potential threat to human lives and property damages.

A second study by Dr. Cerveny as WMO Rapporteur (Cerveny et al. 2017) considered two categories of the largest known lightning impacts on people since 1873. The first category is the "highest mortality (indirect strike) associated with lightning" which was accepted as when 469 people were killed in a lightning-caused oil tank fire in Dronka, Egypt, on 2 November 1994. The second category is the "highest mortality directly associated with a single lightning flash" as the lightning flash that killed 21 people in a hut in Manica Tribal Trust Lands, Zimbabwe. Such information is valuable for improved understanding of the situation leading to mass lightning casualty events, and possible impacts of lightning.

7.3.3   Observatories

Due to the frequently low humidity in the atmosphere, Arizona is one of the best places in the U.S. to stargaze. Currently, Arizona has 31 astronomical observatories including many world-renowned facilities such as Lowell Observatory (where Pluto was found), Kitt Peak National Observatory (first to observe an instance of gravitational lensing in the universe), and Steward Observatory (with a reflector known as the "All-American Telescope"). Astronomers/cameras at the observatories sometimes capture lightning phenomena as shown in the lightning photo (Fig. 3.9) that was taken by Gary Ladd on June 4, 1972, at the Kitt Peak Observatory near Tucson.

7.3.4 Middle Atmosphere Electrical Phenomena Observed in Arizona

The weird, naughty, and rare lightning—Red sprites, blue jets, pixies, elves...what do these words have to do with lightning? What are they and where can they be found?

During the last several decades, an entirely new array of middle atmospheric lightning-related phenomena has been found since their early documentation by Sentman and Wescott (1995). They occur over large thunderstorm complexes and have been documented in many areas of the world (Lyons et al. 2003, 2008, 2019, 2022; Lang et al. 2013; Soula et al. 2015; van der Velde 2019). They had been there all along, sometimes attributed in confusion to UFOs, because they are (1) very faint, (2) short-lived, and (3) rare. The underlying cause of these events is usually a large thunderstorm complex, but they can only be seen well at night, and they mainly occur over the tops of large thunderstorms. Given the frequent mesoscale convective systems such as those shown in Figs. 1.15 and 1.16 during the monsoon season over southern Arizona (Holle and Murphy 2015) and northern Mexico (Sect. 4.1), the authors expected that sprites and other optical events were occurring. However, waiting for a sprite to be observed in southern Arizona away from light pollution is not an easy task, since the best viewing areas do not have readily accessible roads. The first known Arizona sprite was captured in 2018 west of Tucson near the small town of Ajo by John Sirlin, one of the Arizona lightning photographers mentioned in Sect. 3.2.2. His photo at https://www.washingtonpost.com/news/capital-weather-gang/wp/2018/07/20/elusive-red-sprites-caught-dancing-above-an-arizona-thunderstorm/ was the result of waiting at the right place and right time with the right equipment!

This new menagerie of lightning types brings us to the end of the story of Arizona lightning up to now. Clearly, a great number of world-class advances in basic and applied knowledge of lightning have taken place within the state, especially around Tucson. Who knows what new projects and people will come to take advantage of the clear skies, storms over mountains, and unique views in Arizona in the future!

7.4   Interviews

In 2014, joint meetings of the International Lightning Detection Conference (ILDC) and International Lightning Meteorology Conference (ILMC) were held in Tucson. At this meeting, one-time awards called "Pioneers of the NLDN" were presented to a small group. Figure 7.7 includes many of the core members of the development of the NLDN. Subsequent to this meeting, we were able to interview three of the primary individuals who made the NLDN possible. In the following interviews, their fascinating first-hand accounts of the early stages of NLDN sensor technology are described, which soon led to the rapid growth of the NLDN to cover the U.S.

7.4.1   Dr. E. Philp Krider

7.4.1.1   Early Background at the University of Arizona

Holle: This chapter is about the history of the NLDN and how we got there, how it happened here. I think what's really interesting is what was the foundational research like Leon Salanave's before the idea came for the lightning

Fig. 7.7 Awardees of the "Pioneers of the NLDN" at the Vaisala-sponsored ILDC/ILMC meetings in Tucson, Arizona in March 2014. Drs. Krider, Cummins, and Orville are interviewed in Sect. 7.4, while Carl Noggle and Dr. Pifer are critical members of the early development of the NLDN (© R. Holle)

detection. There was background research instrumentation development here at the Atmospheric Sciences Department?

Krider: Yes, the Atmospheric Sciences Department was affiliated with the Institute of Atmospheric Physics. Two separate organizations within the University of Arizona. That distinction is becoming obscured as time goes on, but the Institute was started by among others, Dick Kassander, who was an experimental physicist. And so, he obviously was interested in measurements, and he had a program on cloud photometry, and various other things. He had a student here get a Master's degree, I think, who was interested in lightning. Leon Salanave was a research associate at that time, also interested in lightning. Leon had developed a scheme for time-resolved lightning spectroscopy, where he could rotate the film drum of a big aerial camera along the wavelength of a long focal length format and get really good resolution on slitless spectroscopy by virtue of the fact that cloud-to-ground lightning is mostly vertical. And it's a very narrow source. He didn't need a slit and collimator. He could just take a picture of the lightning through a diffraction grating and resolve the different spectrum lines (Salanave 1980). That program had been going on for several years when I came in 1963 as a research student here in physics. I was interested in finding a Master's thesis project and so Dick Kassander guided me to Leon, who I led to a project where I did a time-resolved study of the individual spectrum lines within return strokes. The rotating film drums in those days were so slow. All Leon could do is time-resolve the spectra of individual return strokes within the discharge. So, in effect he was getting time-averaged spectra over the duration of the bright phase of return strokes. And I came in and got some data on the time-resolved spectrum components of individual spectrum lines with photoelectric sensors. I did that until 1965 or thereabouts and wrote a Master's paper on that. And then, I transferred down to the physics department for my Ph.D. and did a research project on the search for quarks and cosmic radiation. Quarks were at that time a hypothetical elementary particle that is a constituent of the protons and neutrons in the nucleus. And we didn't find any quarks that were free and separate from the cores of extensive cosmic showers, but we looked and that was fun.

Anyway, once you get started on some of the lightning stuff, as you know, well, it's hard to give it up. And so, I continued to do measurements of photoelectric output from return strokes and other processes in collaboration with Martin Uman, who was an assistant or associate professor of electrical engineering at that time and interested in plasma physics and wrote a book on plasma physics. And he was very interested in the spectroscopy. So, Uman got interested in the spectroscopy about the time I was finishing my Master's. So, it's unfortunate because I could have maybe done a Ph.D. with Uman instead of transferring back to physics. Physics was my original field, and I was very happy down there and had a very very good research experience in the Physics Department. And I continued to maintain an interest in lightning and made measurements of the photoelectric output from lightning. How bright lightning was...a topic which I'm still interested in and I'm working on it with Mason Quick (Quick and Krider 2013).

And as time went on, I maintained an interest in lightning and long sparks. I did some experiments with Martin and George Dawson, another professor in the Atmospheric Science Department. George and I went up to the Westinghouse Research Labs after Martin Uman transferred back there in the late 60s. And he did some experiments on long sparks, so we did comparisons between lightning and long sparks spectroscopic comparisons, and how much light comes out comparisons. And at that point, I'm concentrated almost exclusively on my physics thesis until about 1971. Then I got an NAS-NRC postdoc at the Manned Spacecraft Center in Houston right at the tail end of the landing on the moon.

Balloons!

Krider: I arrived in Houston and was working at the Manned Spacecraft Center in June of 1969, which was just a few months, a few weeks literally after the Apollo 11 astronauts landed on the moon. And I worked in the space physics section there doing high-altitude studies of cosmic rays, measuring the energy spectrum of the primary cosmic rays using high-altitude balloons. We flew a 25,000-pound payload to a height of about 130,000 feet. I think we still have the heavy payload ballooning record.

7.4.1.2 Transition to Kennedy Space Center Studies

Krider: In November of 1969, Apollo 12 was launched just after Apollo 11. And Apollo 12 triggered two lightning discharges during launch. The first one was a multi-stroke CG lightning that struck in two different places on the ground and the second one was an IC lightning at about 14,400 feet. The vehicle was at that height. And that was interesting because NASA had never considered the possibility that they would trigger lightning during a launch. And so that led to a whole parallel story of what happened, helping them understand what happened (Godfrey et al. 1970; Uman and Krider 1989).

Holle: And you were there already?

Krider: I wound up being the internal guy that coordinated. I knew every-body in the field of lightning who was still active at that time. So, I invited most of them to come to Houston and help provide advice to NASA on what happened and how to prevent that from happening in the future. We wrote a report, an internal NASA report, that describes the incident and made recom-mendations for launch rules (Godfrey et al. 1970). But anyway, that whole Apollo experience led to a long-term involvement with KSC in lightning.

Then in 71, I think it was, the Apollo program was cut back. They canceled the last three missions and decided not to go back to the Moon after Apollo 17 and scrapped the rest of the hardware they had already developed. And about that time, Martin had developed a new theory; Martin and McClain had developed a theory for the production of electromagnetic fields by light-ning radiation fields. And I helped him program the results of that theory, and we published a paper describing the theory and the results of the computer calculations in the *American Journal of Physics* in about 75 or 76 (Krider et al. 1975). And so, I became really interested in what those fields look like during the return stroke phase particularly. Martin had a student at the University of Pittsburgh who developed an antenna for measuring the broadband elec-tric field with a flat plate antenna. It sat up on a 1-m-high post and had an integrator and follower mounted right under the plate and they drove a long cable with the signal back to the central station and were measuring the electric fields. His measurements tended to validate the Uman model for the production of those fields, which was exciting. So, I got very interested in that after programming the theory.

In any case then in the mid-seventies I guess, the question was, well that's the electric field waveform; what's the magnetic field wave form? And so, then we just started experimenting with loop antennas. Carl Noggle got involved with loop antennas and he's of course very good experimentally and he hadn't actually used operational amplifiers, and I had at the Houston balloon flight. We had a lot of op amps in those days, and I introduced Carl to op amps which was kind of fun, because he took that straight to the top of the field. And at that time, the solid-state technology was coming along real fast, partic-ularly in the analog circuit regime. And the analog circuits were the two that were particularly valuable were things like the old 141. Very simple op amp, idiot-proof that works pretty well. Doesn't have quite the bandwidth you need to do the lightning detection right. But it was very inexpensive, less than a dollar a chip, and made life a whole lot easier. The other thing that we used a lot of were the nonlinear switches, analog switches like a DG201. Those had just come out and you could gate an analog signal through those switches and filter out low-amplitude noise very easily. So, we took advantage of this solid-state circuit technology for both analog circuits and later for digital circuits and developing lightning detection.

Lightning Measurement Firsts

Krider: We had probably made some of the very first magnetic field measurements of lightning with enough bandwidth to see what was really there. And we used the results of those measurements in developing the lightning sensors. And all that work was done here in Arizona and in Florida at KSC. KSC sponsored a lot of our earlier work, as did the Office of Naval Research in Washington.

7.4.1.3   Early Forest Fire Detection

Holle: So, the electronics came along just at the right time?

Krider: Yes, it was parallel, as soon as we got a new electronic circuit, we figured out a way to do this. The only thing that wasn't developed highly at that point was things like GPS, which is really the heart of the multiple station idea.

Holle: Communications were not so good?

Krider: Yes, communication technology in general was pretty poor.

300 baud

Krider: We had trouble getting a 300 baud phone line and the first lightning detectors were really single station detectors that we deployed in Alaska and Canada for detecting forest fires.

With a single station, if you know the direction to the lightning, you could tell approximately how far it was from the amplitude of the signals. And so, we had a way of plotting the amplitudes of the signals on a paper chart recorder. We would either plot the amplitude directly where the little ones are the ones that are far away, and the big vectors are the ones that are up close, or we could invert the signal and do it the other way where we made the little signals big. We put a vector out on paper and then we would plot a point and then come back to the origin of this on a plotter (Sect. 6.5). And when you come back, we came back logarithmically, so the amplitude of the signal is really a log amplitude, and you could tell how big it was either from the tip that was plotted or most generally usually from the tail where the pen lifted up off of the paper.

First Application for Fire Detection

Krider: And that was actually a useful tool for them because they could vector the airplanes that go out and look for fires and they knew about how far away, and then they could tell the exact direction. Then they look for the fire.

Holle: The first Alaska ones were single station?

Krider: Yes, they were all single stations in the Northwest Territories, Northern Alberta, places like that, where it was really remote, but they were still very useful for the firefighters. The earliest two-station that actually worked was I think in Ontario. There's a real nice fellow up there Trevor Woods, made a lot of important contributions as a user and you were working on the Florida FACE one, I think. And the guy was Albert Boulanger was the technical guy.

Holle: Right, and Mike Maier. So, you built those sensors here?

Krider: They were all built here. And eventually, well, we started to form a company—Lightning Location and Protection Inc. that the manufacturing part was based here, and the consulting part was based in Florida with Martin Uman. But all the manufacturing went on here.

Holle: I have a story I want to ask you about. I asked Martin Uman some years ago, it must have been at the ILDC. How did you and Martin develop the criteria for the waveform detection? He told me that … and I repeated this last year and he said that he does not remember saying this. Anyway, he said you, the two of you were at KSC one summer walking down the beach in the evening after looking at data all day. And one of you said every time there's a CG flash, there's a negative excursion and then one of you said, but then it goes and overshoots, goes up and one of you said, but then it comes back down, and so on and he said those are the five criteria needed to detect lightning. Remember some conversation vaguely like this?

Krider: Not so much with Martin.

Holle: He denied having told me that, but I remember that very clearly.

Krider: When we had those single station direction finders, Carl (Noggle) and I used to operate it up in the penthouse of the old Physics building over there, up there where that glassed-in penthouse where you have good visibility back in those days, good visibility all around. Carl had developed a way of using the horizontal sync pulse video signal. We were experimenting with ways of getting timing. And one way of getting good multiple station timing within the range of a television transmitter is syncing on the horizontal sync pulse of the video signal, because those come very very precisely, and you can pick them up anywhere you can pick up a television signal. And we were experimenting with synchronizing multiple stations around the Tucson area using those sync pulses. That was before the GPS receivers came out in a very compact cheap format, and you could buy a GPS receiver starting in about the early or mid-80s for $10,000. But clearly that's too expensive to run on a lightning sensor. So, we were experimenting with other ways of getting multiple stations synchronizing. That was an obvious way. And so, Carl had synchronized a television display to the output of a single station direction finder, so we could plot what the television saw at the same time we plotted the vectors. And when you looked at the display of the magnetic direction on top of the image, you can really tell what it is that's making the pulse.

Xerox Machine Interference

Krider: And so, we can see what kind of pulses made the return strokes and what kind of pulses made other things, and the other things you didn't want obviously depending on a lot of what the source was—if it was a Xerox machine, which was a frequent problem. Xerox machines make pulses that look exactly like return strokes.

Holle: Oh really?

Krider: And the pulse gets into the power system and then that couples into your electronics. So, I think I remember what Martin may have remembered. We knew what the return strokes looked like, and we knew how to gate on the peak of return strokes, which was the real secret of our detecting a known process. Return strokes at the right time within the process, the peak, to get very accurate magnetic directions within the first 10 or 15 microseconds if you have a very precise direction. And the problem was that other stuff looked like return strokes, like Xerox machines for example. And so, then we took advantage of pulse shape discrimination to eliminate practically all the background noise. And that was what I might have had an idea on how to do that, because when I searched for quarks in the cosmic radiation, you get lots of transient signals coming through your cosmic ray telescope, but you can always tell what's a cosmic ray versus other kinds of noise from the shape of the signal. And I was looking for quarks which have a unique shape; they were thought to be only fractionally charged at that time. And we had a rise time criterion and a width criterion and other things in our scintillation counter. And we had six big 2-m square, actually, eight 2-m square scintillation counters stacked floor to ceiling that was acting as our Cosmic

Ray telescope. Also, we had big wide gap spark chambers embedded in that array to measure the direction of the cosmic rays, because you need to make a geometrical correction of the signals.

Quarks to Lightning

Krider: So, I had a rise time and the width as the two main criteria in the quark experiment and I said, well, we can do the same thing on lightning.

And with the leader pulses of the polarity, the leader coming into the return stroke tends to be almost the same as the return stroke. Sometimes they're flipped over but… So, we had a requirement in those early shape criteria that the leaders coming in before the first return stroke be the same polarity as the return stroke that followed. And then, you tended to not have a lot of bipolar leader pulses in the few hundred microseconds before the return strokes. So, we can have a bipolar reject criterion before the return stroke, and we had a requirement that the leader pulses with the same polarity of the return stroke and then the rise time in the width and the second peak following the first peak, had to be lower than the first peak. Those were all about the PTK.

Holle: The five I remember.

Krider: Yes, the PTK, the pre-trigger kibosh was the leader polarity pulse. The polarity of the leader pulses being the same as the return stroke, the absence of an opposite polarity before the return stroke, the rise time width, oh, and when you see a cloud discharge, it tends to have a lot of bursts, lots of pulses. So, we rejected bursts like that.

Holle: So, these parameters and criteria were in the single station Alaska sensors right away?

Krider: Well, in 76 I think, we went up to Alaska and installed five or six stations and found that they really didn't work very well primarily because of the interference from background stuff. Had a lot of trouble with their Xerox machines. And so, for the best test, I had them turn the Xerox machines off while we were expecting lightning, and then, in about…it was in 76 and 77, we really implemented most of the shape criteria and we tried them up in Alaska. They didn't have any communications in the early days. And the communications they did have really weren't very good. They really worked very well, particularly during the lightning storms. So that was how that all got started.

7.4.1.4 Growth of LLP

Krider: Then in the early 80s, I withdrew from the effort. We sold the commercial Lightning and Location and Protection business to a technology transfer outfit at MIT—Dynatech (Sect. 6.4). Martin Uman began here in the electrical engineering program and then left in the late 70s, went to Westinghouse Research Labs in Pittsburgh.

Holle: Was it 60s or 70s?

Krider: It would have been late 60s. He wrote a book at Westinghouse. His first couple of lightning books, *Understanding Lightning* (Uman 1971) was his first, and his second one called *The Lightning Discharge* (Uman 1987) primarily when he was at Westinghouse. And so, he went to Westinghouse in 67 or 68, because we were doing photoelectric experiments on long sparks in the late 60s.

Holle: But after the Apollo, the two of you…

Krider: That was all before.

Krider: He was on the committee I brought in (Godfrey et al. 1970). They were Marx Brook, Martin Uman, Charlie Moore, Ted Pierce, Don Fitzgerald. Who's that guy from Minnesota? George Fryer. Those are the principals, and did I mention the aircraft guy Don Fitzgerald? So, with Moore, Brooke, and Fitzgerald, we had aircraft measurements covered pretty well. Ted Pierce also knew a lot about that stuff (Pierce 1976). Uman and I knew a little bit about lightning. Pierce had done a lot of work on lightning, and he was in England. So, then we had a meeting at the AGU fall of 69 where we presented all our initial thoughts on what happened and what to do about it. And then from that point, it just sort of persisted as a topic of considerable interest, triggered lightning started up. Then eventually the artificial lightning, triggered lightning started up and then that all led to a program a lot of places that the French got involved. But anyway, all of our work we did here; the Vaisala stuff, the hardware. Uman ran a consulting component of our initial business; he got work with Electric Power Research Institute (EPRI) people and some of the aviation people—McDonnell Douglas at St. Louis and some others. He pulled together studies of the characteristics of lightning radiation fields primarily for the aircraft industry. But then he got a program going for the electric power research interests in Tampa Bay. That was really one of the first, one of the first two or three position analyzers went into that Tampa Bay experiment. And that was headed up primarily by Mat Darveniza; he was in Australia, who is a very good guy (Darveniza and Uman, 1979, 1982).

7.4.1.5 Die Blitz-Bibel

Holle: Daile and I have looked at least three or four times through Die Blitz-Bibel (Fig. 7.1).

Krider: What?

Holle: The Lightning Bibles, the blue notebook that you've collected. These are Die Blitz-Bibel. Remember that?

Krider: I'd have to see it. Leon put them together. No, they probably ended about Leon's time, the early 70s. I came to the University of Arizona when Leon left. Leon left to take up the job as director of the Astronomical Society of the Pacific in the beginning in 1971 and I arrived here in April and picked up the program.

Holle: It was back when bibliographies were hard to come by.

Krider: Yes, and he was very interested in all those early papers. Or he got them from a guy named Jensen; Merle Jensen from I think it was University of Nebraska, a very early lightning researcher. He brought a lot of his papers and works to Leon and got Leon started; some of Jensen's papers were interesting. Jensen did experiments looking for ball lightning and things like that.

Holle: That was in there. In there is a glass slide of a lightning strike from 1915. (Authors' note: The slide was subsequently found to be from Canada not Arizona).

Krider: That would have been from Leon's collection.

Holle: That's where we want to go backwards, to see what the foundational steps are.

Krider: That was the first. Well, before Leon there was some interest in lightning spectroscopy on the part of Kitt Peak. There's a guy named Wallace that had photographic plates making spectra of the stars. They got fogged any time there was a lightning storm in the spectrographs. They were running slit spectrographs and they had a lot of lightning images by accident. And Wallace analyzed a lot of those for what was there; he found mostly molecular emissions which was indicating the kind of things you get when you average over time everything in the lightning storm. It was published in a paper in the early 60s, I think.

Holle: I think it's in this blue book.

Krider: And there was also some very early spectroscopy around the turn of the century. Around 1900, I think, some early spectroscopy in Europe, some Germans or something made from spectra.

Holle: It looks like they did something at Lowell Observatory.

Krider: I think I gave some of that to you. So, check the Montreux conference proceedings, the International Conference on Atmospheric Electricity in 1963, if I remember right. That was some of Leon's earliest work. Leon himself got started as an astronomer doing the site survey for the location of Kitt Peak Observatory. And he came down here near Southern Arizona and camped out on mountain tops for a summer making visibility studies, wind measurements, meteorological measurements, wind, visibility, cloud obscuration, things like that that the astronomers were interested in. He spent one or two years on the mountain tops around here. And you don't have to spend much time on a mountain in Southern Arizona to get a lot of exposure to lightning and he then realized that there hadn't been much work done on lightning and started the spectroscopy studies with Jim Hughes' support.

Holle: That's the link that I didn't know.

7.4.1.6   Early Sensor Technology

Krider: Large scale integrated circuits were just coming out, both analog and digital. The analog we exploited very early through those DG201. Those were an analog basically wide band, you can get anything through them, but only when you gated on it. So that was really nice for our application. Op-amps 741, if it weren't for cheap op-amps, we would have really been hurting.

Holle: Ten years earlier, as I understood it, but you probably couldn't have done this.

Krider: We couldn't have done it nearly as easily as we did. I remember when our first communications were with those voice modems. Dial up, plug into the receiver, wait for the tone and hang it up. And you had 300 baud over those. With some difficulty…if you were up north in Alaska, you couldn't even get 300 baud and in Canada you didn't even have it.

Holle: So, the understanding of the wave shape and so on came about the same time, as the ability to do it led to making the network real time?

Krider: You could do what we did in the direction finders. For quite a while, I mean we had done a lot in the cosmic ray area. One of my early students gated on a radar. I had him gate, a wideband, you know when radar signals come back, they're real broadband, and then you demodulate, you're averaging and stuff like that. What you can gate on the raw radar signal, you can look for lightning signals on the radar return. So, you could do it, but it was hard. I mean it was expensive. We were doing wave shape discrimination with the cosmic ray experiment, running wide gap spark chambers in close proximity throughout the 60s. But it was hard.

7.4.1.7 Network Applications

Holle: When you and Martin and Carl developed the direction finder and the ability to link them together, did you realize how big the applications would become or was it fairly obvious?

Krider: Pretty obvious. We had, you know, a lot of history before us. We knew the work of time of arrival and hyperbolic methods and all that. Those were all well-known applications during WWII. Watson-Watt and those people in England had an application, and I had written Ted Pierce who worked at Cambridge about prior art for the patent. And Pierce wrote me a letter and said that other people had looked at the return stroke, but nobody had gated in the early part of the return stroke before, which was the secret. That was very effective; you got polarization errors coming in after about five or ten microseconds, so that broadened the angle distribution which made it unworkable. The TOA (Time-of-Arrival) was a supplement that came in when the GPS receivers got cheap. And that was early late 80s, early 90s. Prior to that, those GPS receivers were really expensive, 10 or 15 thousand dollars each. And that was in the days when that was real money. So, but now these cheap things, you know, even a cell phone has a GPS receiver in it. So that really brought that alive, but we had read all that Lewis stuff in the patents that originally quoted a lot of that literature.

Holle: But you had an idea that locating lightning in real time was going to be a big deal?

Krider: Oh that! Phenomenologically. Yes, that had never been done before. Nobody paid any attention to that before in a systematic way. The English had done the most, and in our experience, but they're the only ones. The French had done a little. The French had a couple pretty good people doing it during World War I; they were handicapped because their country had been captured by the Germans. And the Chinese…Charlie Moore actually had written that there were direction finders in China in World War II, looking for lightning, but he told me it was a really a tough time to be in China if you were an American. The Germans had a whole spherics group; they were okay, but eventually Hans Volland wrote a couple books and handbooks on lightning. And he had an idea that you could use time-of-arrival to look at the VLF and the ELF impulses, the time delay between the VLF and the ELF, and propagation theory at low frequencies, to figure out the range. But the problem was he didn't know the source. There is enough variability in the source that you can't get a very accurate answer. And he did his work, I think out of Hamburg. Before him, there was Heinz Kasemir and those people; Heinz was working mainly on proximity fuses for anti-aircraft artillery. You get a big perturbation in the atmospheric electric field when you go near a conducting object at altitude, and he was using that to detonate artillery shells. That's how he got his cylindrical field mills and all of that going.

Holle: I watched a great PBS show recently about Watson-Watt, a two-hour program.

Krider: Yes, on PBS. That was really good. Watson-Watt had a real problem with the bandwidth when he wanted to get the short-range. He took advantage of the ionosphere hops.

Watson-Watt

Krider: I was interested in the Watson-Watt literature when I was starting all of this, and I could trace very clearly through the papers that Watson-Watt published from my own experience how the technology evolved in those very early days of the mid-1930s and even earlier. And then you get up to about 38, 39, lots and lots of papers stop at the onset of World War II and he got involved in radar, and that of course was highly classified (Barfield 1947).

7.4.1.8   Arrival of Cummins

Holle: How did you find Ken Cummins?

Krider: Pat Zumbusch worked with Ken back in Wisconsin wherever Pat was before, then he came to here.

Holle: Really?

Krider: Pat worked for the same company Ken did back in Wisconsin, but they knew each other back there. Ken came just at the transition between the direction finding only and the integration of the pulse time of arrival, so he got here just the right time for that. So anyway, Zumbusch brought Ken Cummins. They worked for Biomation, I think it was, back in Wisconsin, Madison. Ken had worked on hearing aids.

Holle: The signal processing?

Krider: Yes. Digitizing and analog signal. See, the bandwidth of digitizers came along with all the technology; I remember how excited I was when I ran the first Biomation waveform digitizer for getting the pre-trigger behavior. It was about eight megahertz, I think. 805. 5 megahertz 8 bits Biomation model 805, eight bits 5 megahertz, and then the 1010 was 10 megahertz 10 bits. And we were I think the first lightning people to get the pre-trigger behavior. Those technologies would have been very useful ten years earlier.

We still operated it for research. We have a series of papers on the pre-trigger behavior of leaders (Krider and Radda 1975; Krider et al. 1977; Uman et al. 1975).

7.4.1.9 Concluding Remarks

Holle: Now, about Tyler Kranz; he had a Master's on Grand Canyon lightning.

Krider: It's probably just the frequency of storms because of the surface cooling but, I don't know what's going on up there. I thought for a while, you'd have an edge effect, as any source of radiation behind a slit appears at the edges of the slit. It kind of like, you know, just by a diffraction effect. But the Grand Canyon is so big, it's probably just the frequency of storms that goes down. But I still think you might see that edge enhancement. I just found an envelope on top of this box. These look like prints of some of the first direction finding antennas that we operated in Alaska. This is Cathy Barker, one of the Alaska fire detection specialists, who got very interested in the technology. There it is on top of the building where we worked in Fort Wainwright. There, it is out at the McGrath field station (Fig. 6.3). I think that was on the Yukon River. There is the display with the vectors. And in this case, they turned it on. Well, there's a logarithmic display of those vectors. It goes both ways, where it turns on and where it turns off. So, this is an example of the display. These are examples of the early storms. I see this is dated August 1976 (Fig. 6.4). That was when we were up there. So, you guys want those?

Holle: I assure you that I will have gone through everything that you gave me…one page at a time. I'm not done yet. Not even close.

7.4.2 Dr. Kenneth L. Cummins

7.4.2.1 Lightning Origination Over Topography

Cummins (*showing the Arizona lightning density map with elevation overlaid*): The relationship between lightning and localized terrain gradient is as extreme in Arizona as I think it could be anywhere in the world. So, you can take a map of lightning density and plot a map of terrain and put them next to each other. And they're pretty darn similar. Or what I do is I layer the lightning over it, and I'm getting higher and higher resolution and longer and longer datasets, and these little features show up. There was a hot spot in the Catalinas; it's one of the peaks just off the Pusch Ridge to the southeast and it's double the density of anything. So, some of them I find by looking at the mountains and then going back and looking at the density map and finding it and say yes or no in terms of feature size and isolation. In other cases, I look at the density map when I find a hotspot and then I go to the terrain map on Google Earth and find that, okay, this is a peak and it's only 5% taller than the one that's two and a half kilometers away from it or a ridge that's two and a half kilometers, but there's a gap in between.

Attractive Distance of Lightning

Cummins: So, it's showing us the attractive distance of lightning. Because if it were bigger, it defines the separation distance of objects being seen by descending channels in a flash, and how far they'll go. If they came from this big (lower resolution), then certain small features would get smoothed out and not show up. But if it comes from something this big (higher resolution), this area over which it grabs is smaller. So, it's a really interesting exercise in terrain effects, and Southern Arizona is just a magical place for them.

Holle: What about Mount Pinacate down in Mexico, just across the border from Yuma? That's an isolated, volcanic peak down there. During the monsoon, I think I've seen it light up.

Cummins: Well, I'm going to bring up my most recent Arizona map, and you can look at it. But there's convergence. You remember Joe Zehnder, looking at the Catalinas, looking at convergence over the Catalinas? I mean mountains make a lot of convergence and they don't have to be huge to flap the butterfly's wings in a certain spot. They're finding convective oddities in areas with just little rolling hills. You probably know it, but some of these sort of spiraling things that then under the right conditions can create local convection and suppression. They're not very tall; they're hundreds of meters, but they're at the right spacing to create the rolls.

7.4.2.2   Prior to LLP

Holle: So Pat Zumbusch hired you? And you were working with him…

Cummins: Before that in Madison, Wisconsin.

Holle: I didn't know that.

Cummins: I was a neuroscientist from 1974 to 1989. I have maybe 30 published papers and a couple of patents on neurodiagnostic procedures and path of physiology of nerves. Then I went to Madison to go to work for a company that was making new medical diagnostic tests. And they were interested in it because I'm an engineer and a physicist. So, they said, well, you could help us develop these new things, and I did for a number of years; that's where I met Pat Zumbusch. He was a controller or a corporate controller, financial person for the company. And two things. One, we were on a bowling league together. Late in my career there, he had moved from corporate controller to a business development person, and they were looking to buy a company and he knew that I had all the background in the engineering and the physics and such to do an assessment. So, we worked together on an assessment of this company and actually ended up recommending that the company, our parent company, did not buy them. But we worked together over a period of about a month or so on that, in addition to bowling and then playing pool and drinking beer together. So, then he came out here and took this job and he got to know the people that were in the company and said, we need somebody like Ken, some different kind of person. And I know he doesn't know anything about lightning, but he can learn that.

Move to Tucson

Cummins: And I was looking to leave Wisconsin—we'd had enough of the cold weather. We had done all the work we were going to do to develop new instruments, and the next thing on the agenda was to build cheaper instruments. So, throwing away features and finding a way to make them cheaper. And I said, I'm not good at that; I can make it more expensive!

Holle: That's not as much of a challenge.

Cummins: I didn't like winter sports and we decided we would look west and south. So, I was out nosing around, and Pat heard that I was. The parent company for LLP at the time was Dynatech, and they had another office in Madison, Wisconsin where we were—Pat left the company and went to work for them. And the job he took was to become president of LLP. So, we needed a real management background and science and technology; that's what Pat had in mind. So, I was the 19th employee, and when the company was sold, we had 134. So, Pat hired me, brought me out here, and Suzanne (authors' note: Ken's wife) and I came out and interviewed and I kind of liked it. Suzanne came out and she really liked Arizona. So, we said, okay, we'll do it.

7.4.2.3   Technology Breakthroughs at LLP

Holle: So, what was it, some of the technology breakthroughs while you were here? Was it the digital technologies that exploded right around that time?

Cummins: It did. What got LLP started and able to make a commercial sensor, even around 1980, was the development of integrated circuits. Things that put more than just a transistor or a couple of transistors into a package. Because you needed to have good things called operational amplifiers that can do good analog signal conditioning, and you can build them in circuits fairly easily without having to have a Hughes Aerospace grade electronic engineer, just a good quality electronic engineer. They could make analog circuits that could amplify things and make a good signal. And then you needed to have a lot of, you have to convert it to digital and then do a lot of processing with that.

Advancements in Technology

Cummins: And to do that, you had to have microcontrollers and both of them grew up together around 1980, from the late seventies to the mid-eighties. And without those, you would have never had the first 80–02 sensor and then it kept evolving. But the next thing that you needed, was a breakthrough that occurred the second year that I was here. And that's the initial development of GPS.

A few of the satellites got put up, they made these giant GPS receivers that were the sizes of a loaf of bread or a half a loaf of bread. They were extremely expensive, but we knew that they were going to get cheaper. So, we said, okay, this is going to take us to the next step. Others had tried at State University of New York and others have tried to make things out of lesser timing references like Loran or the satellite-based stuff, and it was never quite right. But GPS, we knew that there was such a pull and a demand for it that not only was it going to be technically adequate to get the timing that you want, submicrosecond or microsecond level timing, but that it was going to become cheap in a short amount of time because they were putting it up there with the anticipation of global use, and the good quality data was limited to the military early on, but there was just all the word on the street was this is going to go everywhere. And when it does, then things that are this big and cost $5,000 are going to get this big and cost a hundred dollars or fit inside my watch and your phone. So, I got there in 89 and mid-1990, we built our first prototype in 91. We deployed a collection of them to Florida for testing in late 1990. And then, somewhere around 92 or so we built a new replacement sensor that had it integrated into it and populated the western United States with those.

Zhang: The IMPACT? (Authors' note: IMProved Accuracy from Combined Technology—a lightning locating method that combines Time-of-Arrival (TOA) and Magnetic Direction Finding (MDF), as described in Chap. 6 and Cummins et al. 1998).

Cummins: No, that was before the companies merged. So, we did that so that we could merge with the State University of New York and the eastern U.S. and limp along with the Oklahoma lightning network in a little spot in Oklahoma and Kansas, and then the western U.S., where's these upgraded sensors to make the first integrated with time of arrival. And now as that was going on, we had funding from EPRI to do this because the electric utilities wanted it and they, they wanted the NLDN to be commercialized (Sect. 6.4). That's when we merged with Rodney Bent's company and then said, well, we're not going to spend the money to build all these new kinds of sensors. He already has sensors with some time-of-arrival in them, and we're putting GPS like we were at the same time. So, he said, we'll make this mixed network so that we don't have to buy 130 new sensors. We'll leave a lot of them in place, upgrade their clocks to be a better version, a contemporary version, we'll move some of them around and build a network that was better than a demonstration NLDN that we had in 1989.

Cummins: So, SUNYA (State University of New York—Albany) worked hard to try and get that going. We put the last leg of it in, in 90 or so, but we didn't have timing in the western U.S. yet.

Holle: I remember it was angles.

Cummins: So that all grew together between basically 91 and 95 or late 94. Took the NLDN from, from an okay meteorological tool to something that the electric utilities could use. I think in meteorology, you guys could have lived with a couple of kilometer location accuracy.

Holle: That's what they're doing even now. It's still mostly the resolution of the radar and so on is on that size.

Cummins: And you don't really need to know operationally, whether that lightning discharge came out of the center of the convective core or a little off to the side, the storm's going to be a lot bigger than the uncertainty and most of the lightning positions.

Holle: Yes.

Cummins: But the weather folks, we're not going to pay for the cost of the big network. You were not going to buy enough stuff to say, okay, great. This company can keep growing and make money and pay back the electric utility company industry, because they didn't have to. So, we had to have a market and that market required accuracy.

7.4.2.4   Arrival at LLP

Holle: So, you came in eighty-when?

Cummins: 1989.

Holle: 89. Did you see all this coming? In terms of...

Cummins: I knew nothing when I walked in the door. When I walked in the door, I went into a room, a little conference room, and one of the cylinders out at the Vaisala building, there was a small one. I think now they're opened up a little bit, but at one time there were just little ovals called echo chambers. And if you sit in just the wrong spot, your voice gets generated throughout the room, or you can hear voices simultaneously seeming like they're coming from above you.

Introduction to Lightning

Cummins: Phil Krider sat me down to a grease board about this size and wrote out the full differential equations for electrostatic induction and radiation fields to make sure that I fully understood. I said, Phil, I've never seen any of this before; that's really interesting; I'm sure I'll get to know it, but he just wrote it out for me and was explaining it all to me. And so, I'm a neuroscientist, I don't know, but I'm an engineer; I'll figure it out.

And so, I knew nothing, but I just started sampling things and looking around and saw all the potential. I was the head of research and development, which was at that time four people, and we just started growing. Pat pushed the business, and I pushed the technology and the sensing, because in my neuroscience work, I did electricity and magnetism of the body. So, I took electrode sensors and put them in places and made measurements and wrote models and software for signal processing and statistics and glued that all together to say something about whether the body was diseased or not, or whether it conducted a signal at the right speeds, or whether it had had evidence of demyelination, which is a nerve degeneration.

Holle: You have a Ph.D. from Stanford in signal processing, is that the right term?

Cummins: Digital and statistical signal processing. And I effectively have a master's degree in neurobiology. And I did the first year of medical school anatomy and physiology, not to be a doctor, but to get all that down while I was doing my engineering work. Because I was going to work in the neuroscience field and I did for a number of years, the engineering sits on top of all the applications, so you can just move that to another field, and as long as you're doing the same kind of math and physics and measurement, then you can apply it to a different problem.

Holle: Yes. That was what I thought was the case.

Cummins: And that's why I liked people having engineering skills. That's why I like people that have statistical skills. I made Mason Quick take a time series, and Chris Biagi and Stacy Fleenor (Table 7.1) take time series and signal processing classes. I was supposed to do a class for Daile, because they didn't have the one I wanted, and I did a lecture or two. And then we moved on to other things, but Mason uses his engineering skills all the time. And Chris ended up getting a Ph.D. in electrical engineering, so he is working with Amitabh Nag in Florida now.

7.4.2.5   University of Arizona

Cummins: So, we've talked about the technology, and talked about history. You know how I got involved in it. From Phil, you would have gotten from the beginning all the way up to the forestry application and… Did he say how he decided to sell Dynatech? Did you ask him?

Holle: Not exactly.

Cummins: He was tired. I mean, he was sick and tired of trying to make a business work. And he wanted to go back to what he knew, which was teaching and academics.

Holle: I knew that was the motivation and yes, he was stuck in corner he didn't want to be in.

Cummins: Right.

Holle: I guess the other question we'll have to ask is the University of Arizona connection with LLP and GAI (Global Atmospherics Inc.). The foundations were here and then somewhat drifted away as it became its own company, but there's always been a connection to Phil and the students and so on to keep improving the technology and the understanding. So that's part of the story we want to have.

Cummins: It's stayed all along, and it's been there since. Phil sold it in 84 I think, somewhere around 84 or 85 to Dynatech, because he wanted to come back to academics, but he kept a spot in the building, he always had a spot in the building that the company had. So, you visit and he had some old equipment. And one of the jobs of the people at LLP who had worked under him in the past was always making sure that the lightning processers that he used to process data were kept alive and running, so he could come out there and take some data from a field campaign and play with it and get some results. So, they did keep that running for him.

Advisory Board at LLP

Cummins: And in fact, one of the things I had to do when I first got there was to work with Bill Newman to try and make sure that Phil could get some data. But I think it was also a way for him to always know what was going on. And then Pat Zumbusch met him and knew that he wanted to have Phil as a resource because Phil is full of ideas about what to do and how you could make things better. And so, I was brought in and said one of the things you've got to do is make sure you keep in touch with Dr. Krider and tell them where you're going and get his input. And so, I set up what was called an advisory committee for the company, a science advisory committee. And every year we met with them, and it was Martin Uman and Phil Krider and a couple of other people.

Holle: I was in one or two of those meetings.

Cummins: We had invited people, so we'd have four. So, you came a couple of times. Mike Maier came a couple of times, and we would present where the company was going and then people would make comments and say, that's stupid or that's a great idea. And so that bridge was kept there between academics and the company. And it also built a bridge between me and Uman and Krider institutionally. So that when I was ready to start backing out of the industry because I was just getting tired of working in the industry, I mean, I've always been a scientist that was sort of wanting to do something practical. So, I got tied in with business, but then I pulled back because I wanted to do research. So, it was just natural for me to be interested then in collaborative projects with Phil and say, well, what about this? So, he said, well, I got a grad student who wants to look at that. I said, OK, then I helped with his grad student and slowly creeped in that direction. But the science advisory panel was a really useful thing to keep us on the right track and keep us apprised of new findings in the field. From a research standpoint, we could say that's important to either know and incorporate in our instruments for future measurements, or to use that to help us define parameters that then we would make sure that the instruments would deliver to the scientific community.

Scientific Basis

Cummins: The scientific community wasn't a direct, big financial supporter of the company, but they bought a few instruments and then they would tell people, hey, this is really good. Look what we're doing with this now and then commercial people would say, oh, well, what's a good lightning system? Well, these guys are paying attention to us, the scientists. They're moving forward. They built credibility.

7.4.2.6 History

Cummins: Phil's got the history goes all the way back to Watson-Watt and Hurd.

Holle: There are two books called Die Blitz-Bibel that I had over at Vaisala. Daile took them a couple of weeks ago. So, she's going to work more on that section. Anyway, we have enough to go 160 or 180 pages.

Cummins: Oh. Yes. That's plenty. Well, the lightning attachment to terrain and interaction with terrain in Arizona is an interesting piece. You could come back to lightning safety.

Holle: That'll be in one of the chapters.

Cummins: In the physics of things. And to me, it's pretty darn clear that you could use the lightning data to let people know that you don't stand on the tops of mountains on the sides and cliffs. And if you do, get the heck out of there if you see clouds overhead.

7.4.3 Dr. Richard E. Orville

7.4.3.1 University of Arizona

Holle: You came to Arizona when Phil (Krider) and Martin (Uman) were in school?

Orville: I came to Arizona first in 1961. And Martin Uman came as a faculty member. I got my degree in Princeton in 1958; he got his degree in 1957 in Princeton.

Holle: Did you know each other then?

Orville: No. He was in the Electrical Engineering (EE) department, and I was in the Physics department. So, I graduated in 58 and then I had to go into the service; I was in ROTC (Authors' note: Reserve Officers' Training Corps). I went into the military, then I worked a civilian job for a year, then I went to Johns Hopkins for grad school. In 61, I went to Tucson, Arizona and I started as a graduate student there. Phil Krider came a year later in 1962, and Martin Uman came also in either 1961 or 62 as a faculty member.

Holle: Was he in EE?

Orville: Yes. Martin Uman was in EE. Phil was in Physics, and I was in Atmospheric Physics and Meteorology, and the three of us started working together just by chance.

Holle: How did you meet each other?

Orville: I was hired because of the spectroscopy work that I was doing in Johns Hopkins, so I went to work there for the summer and stayed five years to get my Ph.D. But I didn't intend to stay there at the time. It was just that everything came together. I actually joined Leon Salanave who hired me as a grad student. He needed help and I had done this spectroscopy work in Johns Hopkins. So, they asked me to come out there for the summer to work and I stayed there five years. But my brother Harry was there a year before.

Holle: So, when you and Martin and Phil were together, you didn't discover the real-time lightning detection technology concept right away?

Orville: No.

Holle: It was an evolutionary thing?

Orville: That was developed in the 1970s. I came from Johns Hopkins, Uman came from Princeton as a Ph.D., and Phil Krider got his bachelor's degree at Carleton College. So that all started around 61 or 62, the three of us working. Phil had a graduate assistantship; he was in the Physics-Math-Meteorology (PMM) building. You were there?

Zhang: For four years.

Orville: We had desks that were about ten feet away. He was in the Physics Department, but the same building. Yes, 5th floor, actually it was right across from the men's room, or the restroom. You go up the stairs and went to the 5th floor, turn left, and there was the men's room. I assume that there's a lady's room, too. And that's where the grad students were; Phil's desk was in a cubicle next to me. That's how it all started by accident.

Holle: So how do these studies that you guys were doing there lead to the concept of lightning detection in real time? Were there steps that were breakthroughs?

Orville: That was really Phil. Phil was good at electronics. He really, as I remember, developed, and on the side developed the company, the LLP company.

Holle: Was Carl Noggle there?

Orville: Carl Noggle. Yes.

Holle: Carl Noggle, Leon Byerly...the underlying question is where does the basic research that led to the real-time network begin?

Orville: At that time, it was still a little difficult for faculty members to have a company. Now it's done all the time. But you did consulting on the side and that's how some of these companies started; and that's how LLP started. Martin and Phil did it on the side, but they were very careful. Everybody was because it could be a conflict of interest. In later years, it's gotten much easier; now it's encouraged.

Zhang: There was a regulation for that?

Orville: Probably because faculty members were paid by the state, state universities, and so public money was being used to pay faculty. And so, it was discouraged to do private work on the side.

Holle: During the office hours.

Orville: Yeah, during the office hours.

Holle: I remember that. Like myself in the government, I wasn't allowed to do anything like that. I had to get waivers to write books on the side.

Orville: That's why I always stayed away from starting my own company. I could have; I didn't.

7.4.3.2 Consulting

Orville: Lightning information services. What I did is, for some reason I was getting the real time data. This is jumping ahead, that must have been in the 1980s or so and I was getting the information on lightning after the network had been set up. I started doing consulting for insurance companies and they would send me, apparently people would claim lightning struck their house and submitted an insurance report. The insurance companies, some of them, found out that I could determine whether or not there was lightning in the area, and I would do the report and charge $100 or $120. And I guess it was in New York when they found out I was doing that. It was a conflict of interest, and it was such a simple thing to do.

> **Lightning Report Consulting**
>
> I grew to dislike it because I liked teaching, and I liked doing research. Just the routine of charging $100 a report and, I guess I did that in the eighties for six seven, eight years, and I quit doing it because it was boring.

7.4.3.3   SUNYA Network

Orville: Anyway, everything started in, I'm just trying to think around 80, 81, 81–82. And it was around 83 when we installed three or four sensors around SUNY-Albany to start the network. You remember that? It just got bigger and bigger, but it started and was developed in the New England area, and we got ten sensors. We purchased ten sensors and they cost $25,000 each, and so we then hit that limit around 83, 84. All of this was sponsored by EPRI.

Holle: But the customer was EPRI.

Orville: And over a six-year period, we got $12 million, and that was real money back in 1983. But we got $2 million a year, starting in 83, 83 to 89. In 89, we had purchased 75 sensors. Around 83, 84, we broke through that barrier of ten. Because we essentially threatened to build our own; not me, but Ron Henderson said he could do it, but Pat Zumbusch changed their mind. Well, the analogy I could make is just think of weather forecasting. Could you do weather forecasting in the United States, if you had just ten stations here and ten stations there to cover the United States instead of having one Weather Bureau? I knew it didn't make any sense to have ten.

Holle: So, meteorologists having a gap in a network is death.

Orville: You can't do that. And so that's why I was, that was where my idea came from for having one national network because for the weather you need a national network, I don't care whether it's lightning or rain or thunderstorms or what, you've got to have that. So, there was never a doubt in my mind that we had to do it as one network.

Zhang: But who called it the NLDN first? Who named it the National Lightning Detection Network?

Orville: Oh, I named that, actually, I was going to call it the SUNY national network, but that after about a year, it didn't make sense. One of our first drops went to Norman, Oklahoma, and I visited there, and they had the SUNY lightning network. Only there was no SUNY, it was just a lightning network. They didn't mean to display SUNY because they didn't want to admit they were getting it from somewhere else. So, the NLDN is what I named it.

Holle: Very cleverly named, I should say.

Orville: It's caught on now, but there were some people that would put their name with it, but I never wanted to do that sort of thing.

Orville: Well, the key thing compared to today is that EPRI picked up on what we were doing. And then the power companies of the United States who report to EPRI, their headquarters is in the San Francisco area, put pressure on EPRI because it was their money. Power companies in the United States collectively wanted that network to cover the United States. There already was a BLM (Bureau of Land Management) network which covered one third of the country, and that was the western part. But EPRI then funded us for about $2 million a year. I never wrote a proposal. It was more or less, what do you want to do?

7.4.3.4 Alaska Network

Holle: BLM, Alaska, I think BLM Alaska was one of the very first customers.

Orville: BLM. I didn't know the Alaska was part of it.

Holle: But the story I've heard from Phil is that those first two antennas, that we were talking about today, we had in the program in Florida in the late seventies. The next customer was almost immediately BLM Alaska. Lonnie Brown, was it?

Orville: Lonnie Brown, yes!

Holle: He wanted to cover it because they were having a terrible time since the area is so large. They couldn't monitor what was going on. I think what Phil said, years ago, is that I can't build 50 sensors by next January in my office: there's no room for this. We need an office someplace here in Tucson.

7.4.3.5 Early NLDN

Zhang: What was the detection efficiency when the first NLDN was built, the year, in 1970s?

Orville: The kind of numbers we were using, not necessarily substantiated by any measurements, was like 70, 75%. The first year that we got data, we got 13 million flashes. I still bring it up; 13.4 million for the United States, meaning, the 48 states. Of course, the network's improved.

Holle: It's about 21 million CG negatives, and plus whatever the positives are, a couple million, something like that.

Orville: Now the whole thing's running 25 to 30 million, not over 30 million, but in the mid-twenties. But 13.4 million was for 1989 I think that was the first year for the 48 states.

Holle: Positives didn't come around until...

Orville: Ah, there's an interesting story on that. I said how clever Ron Henderson was. In March, in January, we decided to put the network together, so we did an installation.

First Positive CG Detection by SUNYA

Orville: We were bringing up three sensors around Albany, New York, and we put the first network in, and Ron and I went out to get data from the first day when we thought it was complete. And there was a storm nearby and Ron Henderson hooked everything up. And then the first lightning signals that came in were positive. And I said, Ron, you reversed the wire.

Holle: Sure, right.

Orville: And I've never seen him then or since get so mad because he's extremely capable, very, very good. He doesn't really make mistakes. And I accused him of, of reversing the wires. And that's the most insulting thing you could say.

Holle: Challenged you were right there, right?

Orville: Oh. He was very upset. That was, it was March of 83, I think.

Holle: I think Carl Noggle had something to do with doing the positive option on it too, but maybe that was more of the commercial part of it. I don't know.

Orville: Well, so I guess one had the capability of picking up polarity.

Zhang: Was it the waveform?

Orville: Yes.

Holle: It was inverted.

Orville: Yeah, so I just said, well, you reversed the wires.

Zhang: But before that you didn't see any lightning that had positive polarity?

Orville: That was the first lightning. That was the first, March of 1983. I remember that. It's the first time with the sensor with a little network of three sensors. We'd picked up lightning, and when he came in and it's like nine out of the first ten or 20 were positive lightning,

Holle: If you could have been to Florida. You would've never had it.

Orville: No, you wouldn't have had it. So, it was March, it was wintertime. And that's when the positives dominate.

Holle: Did you have an idea that there were that many positives?

Orville: We had not, in my own mind, I may not have even heard of positive lightning. Rust and MacGorman were the first to pick that up (MacGorman and Taylor 1989). That's how new it may have been to me. But if I had heard there were positive lightning, you know, it's very small 4, 5% something like that. Things were changing fast then.

Holle: They had a four-sensor network in Oklahoma, sort of diamond shape, and then Bob Maddox wanted a network in Kansas. We added three more sensors, and I was involved in that. So, it made a seven-sensor network, and then you absorbed that into your network. Then you filled in the rest. People had an Oklahoma and Kansas network, the BLM, and yours. There are all these holes in between, so what's the matter with you people? Can't you get together? Something to that effect.

Orville: Actually, it was Fred Mosher at the AMS (Authors' note: American Meteorological Society) meeting in 1987 who posed the question.

Holle: I think the first flash we have in the dataset at Vaisala is January or something of 89. Our archive goes back to 89.

Zhang: How many sensors were in the network, do you remember? Like a hundred or so?

Orville: Well, I bought 75 sensors.

Holle: BLM had 30 something.

Orville: 30, 35, or so. We had a hundred and now it's 106 I think, or 105.

Holle: It's now in the low 110s.

References

Adang TC (1989) Structure and dynamics of the Arizona monsoon boundary. PhD dissertation, University of Arizona, 121 pp

Barfield RH (1947) Statistical plotting methods for radio direction-finding. J Inst Electr Eng 94(Part IIIA):673–675

Biagi CJ, Cummins KL, Kehoe KE et al (2007) National lightning detection network (NLDN) performance in southern Arizona, Texas, and Oklahoma in 2003–2004. J Geophys Res 112:D05208. https://doi.org/10.1029/2006JD007341

Blakeslee RJ, Lang TJ, Koshak WJ et al (2020) Three years of the lightning imaging sensor on board the international space station: Expanded global coverage and enhanced applications. J Geophys Res Atmos 125(16):e2020JD032918

Block K (2011) Fulgurite classification, petrology, and implications for planetary processes. MS thesis, University of Arizona, 68 pp

Campos LZS (2016) On the mechanisms that lead to multiple ground contacts in lightning. PhD dissertation, Instituto Nacional de Pesquisas Espaciais (INPE), 248 pp

Candela Garolera A, Cummins KL, Madsen SF et al (2015) Multiple lightning discharges in wind turbines associated with nearby cloud-to-ground lightning. IEEE Trans Sustain Energy 6(2):526–533

Cecil DJ, Buechler DE, Blakeslee RJ (2015) TRMM LIS climatology of thunderstorm occurrence and conditional lightning flash rates. J Clim 28(16):6536–6547

Cerveny R, Bessemoulin P, Burt CC et al (2017) WMO assessment of weather and climate mortality extremes: lightning, tropical cyclones, tornadoes, and hail. Wea Clim Soc 9:487–497

Cramer JA, Cummins KL (2014) Evaluating location accuracy of lightning location networks using tall towers. In: Preprints of the 23rd international lightning detection conference, Vaisala, Tucson, Arizona, 20–21 Mar 2014

Cummins KL (2021) Global patterns of flash optical properties based on long-term ISS-LIS observations. In: Proceedings of the GLM science meeting, online, 21–23 Sept 2021. https://goes-r.nsstc.nasa.gov/home/meeting-agenda-2021

Cummins KL, Quick MG, Rison W et al (2014a) Overview of the Kansas Windfarm 2013 field program. In: Preprints of the XV international conference on atmospheric electricity, Norman, Oklahoma, 15–20 June 2014a

Cummins KL, Zhang D, Quick MG et al (2014b) Performance of the U.S. NLDN during the Kansas Windfarm 2012 and 2013 field programs. In: Preprints of the 5th international lightning meteorology conference, Vaisala, Tucson, Arizona, 20–21 Mar 2014b

Cummins KL, Murphy MJ, Bardo EA et al (1998) A combined TOA/MDF technology upgrade of the US National Lightning Detection Network. J Geophys Res 103(D8):9035–9044

Darveniza M., Uman MA (1979) Lightning studies of transmission lines. In: Proceedings of the IEEE power engineering society conference on overhead and underground transmission and distribution, Atlanta, Georgia, 01–06 Apr 1979

Darveniza M, Uman MA (1982) Lightning protection of distribution lines. Final report, U.S. Dept. of Energy, DOE/ET/29066–1, Univ. of Florida

Fleenor SA, Biagi CJ, Cummins, KL et al (2009) Characteristics of cloud-to-ground lightning in warm-season thunderstorms in the Central Great Plains. Atmos Res 91:333–352

Godfrey R et al (1970) Analysis of Apollo 12 lightning incident. NASA-TM-X-62894, 87 pp

Goodman SJ, Blakeslee RJ, Koshak WJ et al (2013) The GOES-R geostationary lightning mapper (GLM). Atmos Res 125:34–49

Gressent A, Sauvage B, Defer E et al (2014) Lightning NOx influence on large-scale NOy and O3 plumes observed over the northern mid-latitudes. Tellus B 66, 17 pp

Gungle B, Krider EP (2006) Cloud-to-ground lightning and surface rainfall in warm-season Florida thunderstorms. J Geophys Res 111(D19203)

Hansell SA, Wells WK, Hunten DM (1995) Optical detection of lightning on Venus. Icarus 117:345–351

Holle R, Murphy MJ (2015) Lightning in the North American monsoon: an exploratory climatology. Mon Weather Rev 143:1970–1977

Holle RL, Brooks WA, Cummins KL (2021) Lightning occurrence and casualties in U.S. National Parks. Wea Clim Soc 13:525–540

Jacobson EA, Krider EP (1976) Electrostatic field changes produced by Florida lightning. J Atmos Sci 33:103–117

Jerauld J, Rakov VA, Uman MA et al (2005) An evaluation of the performance characteristics of the U.S. National Lightning Detection Network in Florida using rocket-triggered lightning. J Geophys Res 110(D19)

Jourdain L, Hauglustaine DA (2001) The global distribution of lightning NOx simulated on-line in a general circulation model. Phys Chem Earth Part C Solar Terrestrial Planet Sci 26(8):585–591

Kassander AR, Stewart RM Jr (1955) A simple low-inertia anemometer of the three-cup type. Bull Amer Meteor Soc 36:384–389

Kempf NM, Krider EP (2003) Cloud-to-ground lightning and surface rainfall during the Great Flood of 1993. Mon Weather Rev 131:1140–1149

Koshak WJ, Krider EP (1989) Analysis of electric field changes during active Florida thunderstorms. J Geophys Res 94:1165–1186

Krider EP (1973) Lightning spectroscopy. Nucl Instrum Method 110:411–419

Krider EP, Blakeslee RJ (1985) The electric currents produced by thunderclouds. J Electrostatics 16:369–378

Krider EP, Radda GJ (1975) Radiation field waveforms produced by lightning stepped leaders. J Geophys Res 80:653–2657. https://doi.org/10.1029/JC080i 018p02653

Krider EP, Leteinturier C, Willett JC (1992) Sub-microsecond field variations in natural lightning processes. Res Lett Atmos Electr 12:3–9

Krider EP, Uman MA, McLain DK (1975) The electromagnetic radiation from a finite antenna. Am J Phys 43:33–38

Krider EP, Weidman CD, Noggle RC (1977) The electric field produced by lightning stepped leaders. J Geophys Res 82:951–960. https://doi.org/10.1029/JC0 82i006p00951

Lang TJ, Cummer SA, Rutledge SA et al (2013) The meteorology of negative cloud-to-ground lightning strokes with large charge moment changes: Implications for negative sprites. J Geophys Res Atmos 118:7886–7896

Lang TJ, Pédeboy S, Rison W et al (2017) WMO world record lightning extremes: longest reported flash distance and longest reported flash duration. Bull Am Meteor Soc 98:1153–1168

Leary LA, Ritchie EA (2009) Lightning flash rates as an indicator of tropical cyclone genesis in the Eastern North Pacific. Mon Weather Rev 137:3456–3470

Leteinturier C, Weidman C, Hamelin J (1990) Current and electric field derivatives in triggered lightning return strokes. J Geophys Res 95:811–828

Lorenz RD (2018) Lightning detection on Venus: a critical review. Prog Earth Planet Sci 5:34. https://doi.org/10.1186/s40645-018-0181-x

Lyons WA (2022) Inside the world of sprite chasing. Weatherwise 75(6):14–23

Lyons WA, Nelson TE, Armstrong RA et al (2003) Upward electrical discharges from thunderstorm tops. Bull Am Meteor Soc 84(4):445–454. https://doi.org/ 10.1175/BAMS-84-4-445

Lyons WA, Cummer SA, Stanley MA et al (2008) Supercells and sprites. Bull Am Meteor Soc 89:1165–1174. https://doi.org/10.1175/2008BAMS2439.1

Lyons WA, Bruning EC, Warner TA et al (2019) Megaflashes: just how long can a lightning discharge get? Bull Am Meteor Soc 100:E73–E76. https://doi.org/10. 1175/BAMS-D-19-0033.1

MacGorman DR, Taylor WL (1989) Positive cloud-to-ground lightning detection by a direction-finder network. J Geophys Res 94:13,313–13,318

Mach DM, Blakeslee RJ, Bailey JC et al (2005) Lightning optical pulse statistics from storm overflights during the Altus Cumulus electrification study. Atmos Res 76(1–4):386–401

Maier LM, Krider EP (1986) The charges that are deposited by cloud-to-ground lightning in Florida. J Geophys Res 91:13,275–13,289

Maier LM, Krider EP, Maier MW (1984) Average diurnal variation of summer lightning over the Florida peninsula. Mon Weather Rev 112:1134–1140

Maier LM, Lennon C, Britt T et al (1995) Lightning detection and ranging (LDAR) system performance analysis. In: Proceedings of the international conference on cloud physics. American Meteorological Society, Dallas, Texas, 15–20 Jan 1995

Mallick S, Rakov VA, Hill JD et al (2014) Performance characteristics of the NLDN for return strokes and pulses superimposed on steady currents, based on rocket-triggered lightning data acquired in Florida in 2004–2012. J Geophys Res Atmos 119(7):3825–3856

Medici G, Cummins KL, Cecil DJ et al (2017) The intracloud lightning fraction in the contiguous United States. Mon Weather Rev 145:4481–4499

Merceret FJ, Willett JC, Christian et al (2010) A history of the lightning launch commit criteria and the lightning advisory panel for America's space program. NASA/SP-2010–216283, 250 pp

Murray ND, Orville RE, Willett JC (2005) Multiple pulses in dE/dt and the fine-structure of E during the onset of first return strokes in cloud-to-ocean lightning. Atmos Res 76:445–480. https://doi.org/10.1016/j.atmosres.2004.11.038

Minjarez-Sosa C, Castro CL, Cummins KL et al (2012) Toward development of improved QPE in complex terrain using cloud-to-ground lightning data: a case study for the 2005 monsoon in Southern Arizona. J Hydrometeor 13:1855–1873

Minjarez-Sosa C, Castro CL, Waissmann EJ et al (2017) An improved QPE in complex terrain employing cloud-to-ground lightning occurrences. J Appl Meteor Clim 56:2489–2507

Minjarez-Sosa C, Waissmann J, Castro CL et al (2019) Algorithm for improved QPE over complex terrain using cloud-to-ground lightning occurrences. Atmosphere 10:10 pp

Murphy MJ (1996) The electrification of Florida thunderstorms. PhD dissertation, University of Arizona, 167 pp

Nag A, Mallick S, Rakov VA et al (2011) Evaluation of U.S. National Lightning Detection Network performance characteristics using rocket-triggered lightning data acquired in 2004–2009. J Geophys Res 116(D2)

Oram TD, Garner T, Hoeth B (2005) Use of lightning data for space shuttle and Soyuz re-entry and landing forecasts at Johnson Space Center. In: Preprints of the conference on meteorological applications of lightning data. American Meteorological Society, San Diego, California, 9–13 Jan 2005

Orville RE, Henderson RW (1984) Absolute spectral irradiance measurements of lightning from 375 to 880 nm. J Atmos Sci 41:3180–3187

Peterson M, Liu C (2013). Characteristics of lightning flashes with exceptional illuminated areas, durations, and optical powers and surrounding storm properties in the tropics and inner subtropics. J Geophys Res Atmos 118(20):11,727–11,740

Peterson MJ, Lang TJ, Logan T et al (2022) New WMO certified megaflash lightning extremes for flash distance and duration recorded from space. Bull Am Meteor Soc 103:1243–1247

Piepgrass MV, Krider EP (1982) Lightning and surface rainfall during Florida thunderstorms. J Geophys Res 87:11,193–11,201

Pierce ET (1976) The thunderstorm research international project—1976. Bull Am Meteor Soc 57:1214–1216

Prueitt ML (1963) The excitation temperature of lightning. J Geophys Res 68(3):803–811

Quick MG, Krider EP (2013) Optical power and energy radiated by natural lightning. J Geophys Res 118(4):1868–1879

Quick MG, Krider EP (2015) Optical emission and peak electromagnetic power radiated by negative return strokes in rocket-triggered lightning. J Atmos Solar-Terrestrial Phys 136. https://doi.org/10.1016/j.jastp.2015.06.005

Reynolds SE, Brook M, Gourley MF (1957) Thunderstorm charge separation. J Atmos Sci 14(5):426–436

Salanave LE (1980) Lightning and Its Spectrum: An Atlas of Photographs. University of Arizona Press, Tucson, p 136

Scheftic WD, Cummins KL, Krider EP et al (2008) Wide-area soil moisture estimation using the propagation of lightning generated low-frequency electromagnetic signals. In: Preprints of the 20th international lightning detection conference, Vaisala, Tucson, Arizona, 21–23 Apr 2008

Schultz CJ, Petersen WA, Carey LD (2009) Preliminary development and evaluation of lightning jump algorithms for the real-time detection of severe weather. J Appl Meteor Clim 48(12):2543–2563

Schultz CJ, Petersen WA, Carey LD (2011) Lightning and severe weather: a comparison between total and cloud-to-ground lightning trends. Wea Forecast 26(5):744–755

Sentman DD, Wescott EM (1995) Red sprites and blue jets: thunderstorm-excited optical emissions in the stratosphere, mesosphere, and ionosphere. Phys Plasmas 2:2514–2522. https://doi.org/10.1063/1.871213

Soula S, Mlynarczyk J, Fullekrug M et al (2015) Dancing sprites: detailed analysis of two case studies. J Geophys Res Atmos 122:3173–3192

Stall C, Cummins KL, Krider EP et al (2009) Detecting multiple ground contacts in cloud-to-ground lightning flashes. J Atmos Oceanic Tech 26:2392–2402

Uman MA (1964) The peak temperature of lightning. J Atmos Terres Phys 26(1):123–128

Uman MA (1971) Understanding Lightning. Bek Technical Publications, Carnegie, Pennsylvania, p 166

Uman MA (1987) The Lightning Discharge, vol. 39. International Geophysics Series. . Academic Press, Orlando, Florida, 377 pp

Uman MA, Krider EP (1989) Natural and artificially initiated lightning. Science 246:457–464

Uman MA, Brantley RD, Lin YT et al (1975) Correlated electric and magnetic fields from lightning return strokes. J Geophys Res 80:373–376

Valine WC, Krider EP (2002) Statistics and characteristics of cloud-to-ground lightning with multiple ground contacts. J Geophys Res 107:AAC 8-1–AAC 8-11. https://doi.org/10.1029/2001JD001360

van der Velde J, Montanyà J, López JA et al (2019) Gigantic jet discharges evolve stepwise through the middle atmosphere. Nat Commun. www.nature.com/articles/s41467-019-12

Vonnegut B, Vaughan OH, Brook M (1980) Nighttime/daytime optical survey of lightning and convective phenomena experiment. NASA, NOSL, vol 78261

Warner TA, Cummins KL, Orville RE (2012) Upward lightning observations from towers in Rapid City, South Dakota, and comparison with National Lightning Detection Network data, 2004–2010. J Geophys Res 117(D19)

Weidman CD (1982) The submicrosecond structure of lightning radiation fields. PhD dissertation, University of Arizona, 226 pp

Weidman CD, Krider EP (1986) The amplitude spectra of lightning radiation fields in the interval from 1 to 20 MHz. Radio Sci 21:964–970

Weidman C, Boye A, Crowell L (1989) Lightning spectra in the 850-to 1400-nm near-infrared region. J Geophys Res 94(D11):13249–13257

Williams MA (1986) An analysis of the Voyager images of Jovian lightning. PhD dissertation, University of Arizona, 197 pp

Wilson N, Myers J, Cummins KL et al (2013) Lightning attachment to wind turbines in Central Kansas: video observations, correlation with the NLDN and in-situ peak current measurements. In: Proceedings of the European wind energy association conference, Vienna, Austria, 04–07 Feb 2013

Zhang D (2020) An evaluation study of the international space station lightning imaging sensor. Paper presented online at annual meeting of American Geophysical Union, 01–17 Dec 2020

Zhang D, Cummins KL (2020) Time evolution of satellite-based optical properties in lightning flashes, and its impact on GLM flash detection. J Geophys Res Atmos 125. https://doi.org/10.1029/2019JD032024

Zhang D, Cummins KL, Nag A (2015) Assessment of cloud lightning detection by the U.S. National Lightning Detection Network using video and Lightning Mapping Array observations. In: Preprints of the 7th conference on the meteorological applications of lightning data. American Meteorological Society, Phoenix, Arizona, 04–08 Jan 2015

Zhang D, Cummins KL, Bitzer P et al (2019) Evaluation of the performance characteristics of the lightning imaging sensor. J Atmos Oceanic Tech 36:1015–1031

Zhang D, Cummins KL, Nag A et al (2016) Evaluation of the national lightning detection network upgrade using the lightning imaging sensor. In: Preprints of the 24th international lightning detection conference, Vaisala, San Diego, California, 18–21 Apr 2016

Zhang D, Quick M., Lang T et al (2021) Fusing GEO and LEO lightning observations. Paper presented online at the 2021 AGU annual meeting, 13–17 Dec 2021

Zhu Y, Rakov VA, Tran MD et al (2016) A study of National Lightning Detection Network responses to natural lightning based on ground truth data acquired at LOG with emphasis on cloud discharge activity. J Geophys Res 121(24):14–651

Resources for Further Information

Arizona Books

Green CR, Sellers WD (1964) Arizona climate. University of Arizona Press, Tucson, 503 pp. ISBN-13: 978-0816502769

Holle RL, Zhang D (2017) So you think you know lightning: a collection of electrifying fast facts. Vaisala, Inc., 64 pp. https://www.vaisala.com/en/system/files/documents/Lightning-Booklet.pdf, https://lightningdev.umd.edu/aert/Safety.html

Parham M (2017) 100 years of Tucson weather. A special publication of the Arizona Daily Star, Tucson, Arizona. ISBN 978-0-9882562-8-6

Salanave LE (1980) Lightning and its spectrum: an atlas of photographs. University of Arizona Press, Tucson

Sellers WD, Hill RH (1974) Arizona climate, 1931–1972, revised 2nd edn. University of Arizona Press, Tucson, 616 pp. ISBN-13: 978-0816504664

Wetmore II RH (2010) Thirty years of lightning photography in Southern Arizona. ISBN 10:0982566212, 112 pp

Other Books

Cooper MA, Holle RL (2018) Reducing lightning injuries worldwide. Springer Natural Hazards, New York, 233 pp. https://doi.org/10.1007/978-3-319-77563-0

Cooper MA, Andrews CJ, Holle RL et al (2017) Lightning-related injuries and safety. Chapter 5, Wilderness medicine, 7th edn. Elsevier, Philadelphia, Pennsylvania, P. Auerbach, Editor

Elsom DM (2015) Lightning: nature and culture. Reaktion Books, London, UK, 240 pp

© Springer Nature Switzerland AG 2023

R. L. Holle and D. Zhang, *Flashes of Brilliance*,

https://doi.org/10.1007/978-3-031-19879-3

Gookin J, Morris S (2014) NOLS lightning. Stackpole Books, Mechanicsburg, Pennsylvania, 130 pp ISBN 978-0-8117-1364-1

Rakov VA (2016) Fundamentals of lightning. Cambridge University Press, 257 pp

Rakov VA, Uman MA (2003) Lightning: physics and effects. Cambridge University Press, 687 pp

Uman MA (1969) Lightning. McGraw-Hill, New York, 272 pp

Uman MA (1971) Understanding lightning. Bek Technical Publications, Carnegie, Pennsylvania, 166 pp

Uman MA (1986) All about lightning. Dover Press, 167 pp

Uman MA (1987) The lightning discharge, vol. 39. Academic Press, Intl. Geophysics Series, Orlando, Florida, 377 pp

Websites

Arizona Meteorology

Climate Assessment for the Southwest (CLIMAS). https://climas.arizona.edu/

Flagstaff National Weather Service Forecast Office. https://www.weather.gov/fgz/

Phoenix National Weather Service Forecast Office. https://www.weather.gov/psr/

Tucson National Weather Service Forecast Office. https://www.weather.gov/twc/

University of Arizona, Department of Hydrology and Atmospheric Sciences. https://has.arizona.edu/

University of Arizona campus weather. http://www.atmo.arizona.edu/

Ground-Based Lightning Detection by Vaisala

Global Lightning Dataset 360 (GLD360). www.vaisala.com/en/products/systems/lightning/gld360

Lightning densities by country and regions within countries. https://interactive-lightning-map.vaisala.com/

National Lightning Detection Network (NLDN). www.vaisala.com/en/products/national-lightning-detection-network-nldn

Lightning Safety Associations

African Centres for Lightning and Electromagnetics Network [ACLENet]. https://aclenet.org/

Lightning Strike & Electric Shock International [LSESSI]. https://www.lightning-strike.org/

National Lightning Safety Council. http://lightningsafetycouncil.org/

National Weather Service Lightning Safety. https://www.weather.gov/safety/lightn
  ing-safety
South Asian Lightning Network [SALNet]. https://salnet.asia/

Satellite Lightning Data

Eumetsat Lightning Imager. https://www.eumetsat.int/mtg-lightning-imager
Fengyun 4A Imager. http://fy4.nsmc.org.cn/nsmc/en/theme/FY4A.html
Geostationary Lightning Mappers. https://www.star.nesdis.noaa.gov/goes/
International Space Station-Lightning Imaging Sensor [ISS-LIS]. https://ghrc.nsstc.
  nasa.gov/lightning/data/data_lis_iss.html
University of Maryland Lightning Science. https://lightningdev.umd.edu/aert/